A survey of Cenozoic Volcanism on Mainland Asia

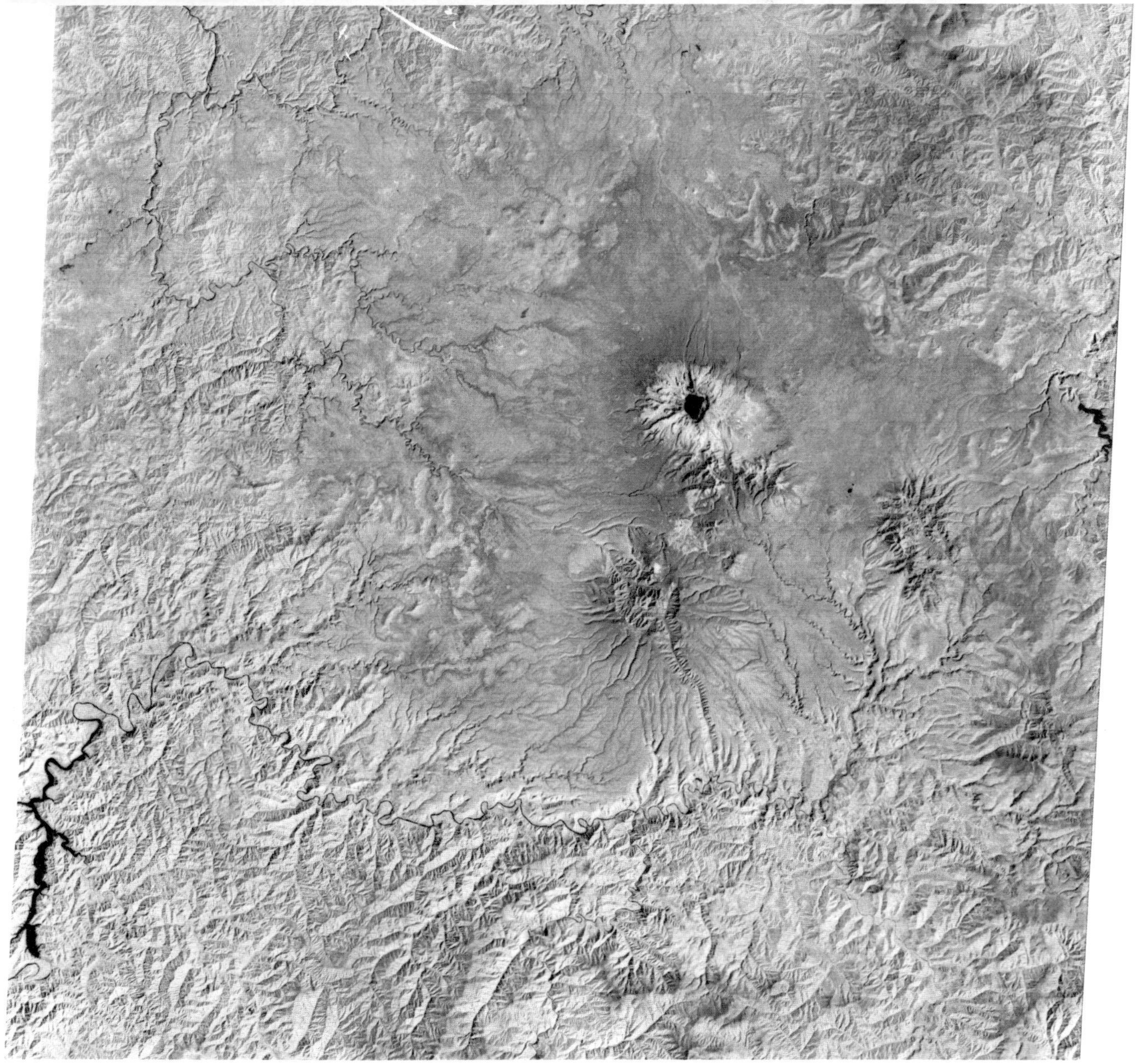

Landsat image of the P'aektu-san volcano.

Cover photo: The round crater on the summit of Nangelaqiushan, about 500 meters across and 40 meters deep, and commonly transformed into a lake in rainy seasons. (From *Wudalianchi Volcanoes in China*, reproduced with permission of Shanghai Scientific and Technical Publishers, Shanghai, People's Republic of China.)

A Survey of Cenozoic Volcanism on Mainland Asia

J. L. Whitford-Stark
Department of Geology
Sul Ross State University
Alpine, Texas 79832

THE GEOLOGICAL SOCIETY OF AMERICA · 1888 ·

SPECIAL PAPER
213

Published by The Geological Society of America, Inc.
3300 Penrose Place, P.O. Box 9140, Boulder, Colorado 80301

GSA Books Science Editor Campbell Craddock

Printed in U.S.A.

Library of Congress Cataloging-in-Publication Data

Whitford-Stark, J. L. (James L.), 1948–
A survey of Cenozoic volcanism on mainland Asia.

(Special paper / the Geological Society of America ;
213)
Bibliography: p.
1. Volcanism—Asia. 2. Geology, Stratigraphic—
Cenozoic. I. Title. II. Series: Special paper
(Geological Society of America) ; 213.
QE527.W48 1987 551.2'1'095 86-31861
ISBN 0-8137-2213-6

Contents

Acknowledgments

I thank Feng Maoseng for providing information on some of the Chinese volcanoes, for translating some technical terms, and for improving interpretations of modern Chinese; Chuck Wood and John Latter for supplying me with data; and Tom Simkin for providing encouragement to do this research. My colleagues J. H. Latter and W. LeMasurier provided thorough reviews that led to substantial improvements in the manuscript. Shannon Rudine provided photographic support. This research has been carried out while I was the recipient of NASA grants NAGW-170, -243, -382, -394, and -605, for which I am extremely grateful.

J. L. Whitford-Stark

Geological Society of America
Special Paper 213
1987

A survey of Cenozoic volcanism on mainland Asia

ABSTRACT

This work presents a catalog of the products of the Cenozoic volcanism on mainland Asia, exclusive of the Koryak and Kamchatka Peninsulas. A synthesis of the available information illustrates that the dominant volcanic landforms are fissures and cinder cones. Some large composite volcanoes are found in Burma, western China, and the northeastern Soviet Union. Peaks of volcanic activity took place during Oligocene-Miocene and Pliocene-Pleistocene time. There is historic evidence for the eruption of fewer than 20 volcanoes; however, many volcanoes occur in isolated, uninhabited regions. In general, with increasing distance from the plate boundaries, there is an increase in alkalinity of the volcanic products, and a tendency for increasing alkalinity with decreasing age. A very small proportion of the rocks plot within the subalkalic field. The principal tectonic controls on volcanism were the early Cenozoic subduction along the eastern seaboard, the collision of the Indian continent, the development of back-arc basins, and interactions between the Asian and North American plates.

INTRODUCTION

The volcanic rocks of central and eastern mainland Asia are poorly known by western geologists, largely for linguistic and political reasons. Additionally, many of the volcanoes are located in extremely isolated areas, many of which have been visited by geologists only within the last few decades. Some areas have been studied only by LANDSAT imagery. I have not visited any of the localities described in the text, and so have had to rely on the work of the writers of the original article and on the capabilities of translators. Where applicable, facts that appear to be in error, or at least stretch the limits of credibility, are emphasized.

Previous compilations, including the *Catalogue of Active Volcanoes of the World* and the *Catalogue of post-Miocene Volcanoes of the World,* both published by the IAVCEI (1951–75; 1973), and *Volcanoes of the World* (Simkin and others, 1981), are limited to descriptions of individual volcanoes rather than to the products of volcanism. The present work attempts to include information on the timing, areal extent, geochemistry, and tectonic controls on Asian volcanism. The area covered by the review comprises that north of the Himalayas and east of 85°E to the Pacific and includes the Soviet Union, Mongolia, China, Burma, Thailand, Kampuchea, Laos, Vietnam, Korea, and the Malay Peninsula. The Kamchatka and Koryak Peninsulas have been omitted as they have already been the subject of many studies.

The review is in two parts: the first part describes the products of Cenozoic volcanism by country and area; the second part discusses the volcanologic and structural implications of Asian volcanism. In Part 1, each area has been assigned a catalog number; these run consecutively, except for numbers 56 and 170, which were late additions. In Part 2, the bracketed number following a location refers to the catalog number in Part 1. The areas that have been cataloged are outlined in Figure 1. An index map indicating the location of text figures is presented in Figure 25. Geographic and rock glossaries are presented in Appendices 1 and 2, respectively. Appendices 3 through 5 list rock types, ages of volcanism, and types of volcanoes indexed to the locality numbers within Part 1. Geochemical analyses are presented in Appendix 7. These analyses are keyed to the locality numbers in Part 1. Appendices 8 and 9 are alphabetical listings of the volcano areas and individual volcanoes within Asia; these are also keyed to the locality numbers of Part 1. Appendix 10 indicates the number of volcanoes within the localities described in Part 1 for which no names were found in the literature.

The catalog does not claim to be comprehensive as there are a number of areas for which I have been unable to find any information. Furthermore, it tends to take several years for articles in non-English journals to be translated or to appear in index volumes. Additional information related to specific areas can be found in the bibliographies of the many references cited herein.

There are various spellings for a large number of the localities described in the text. Additionally, the same locality may be known under a different name in a different country. The published anglicized names herein are used to describe individual localities; they are underlined for emphasis. Alternate names are

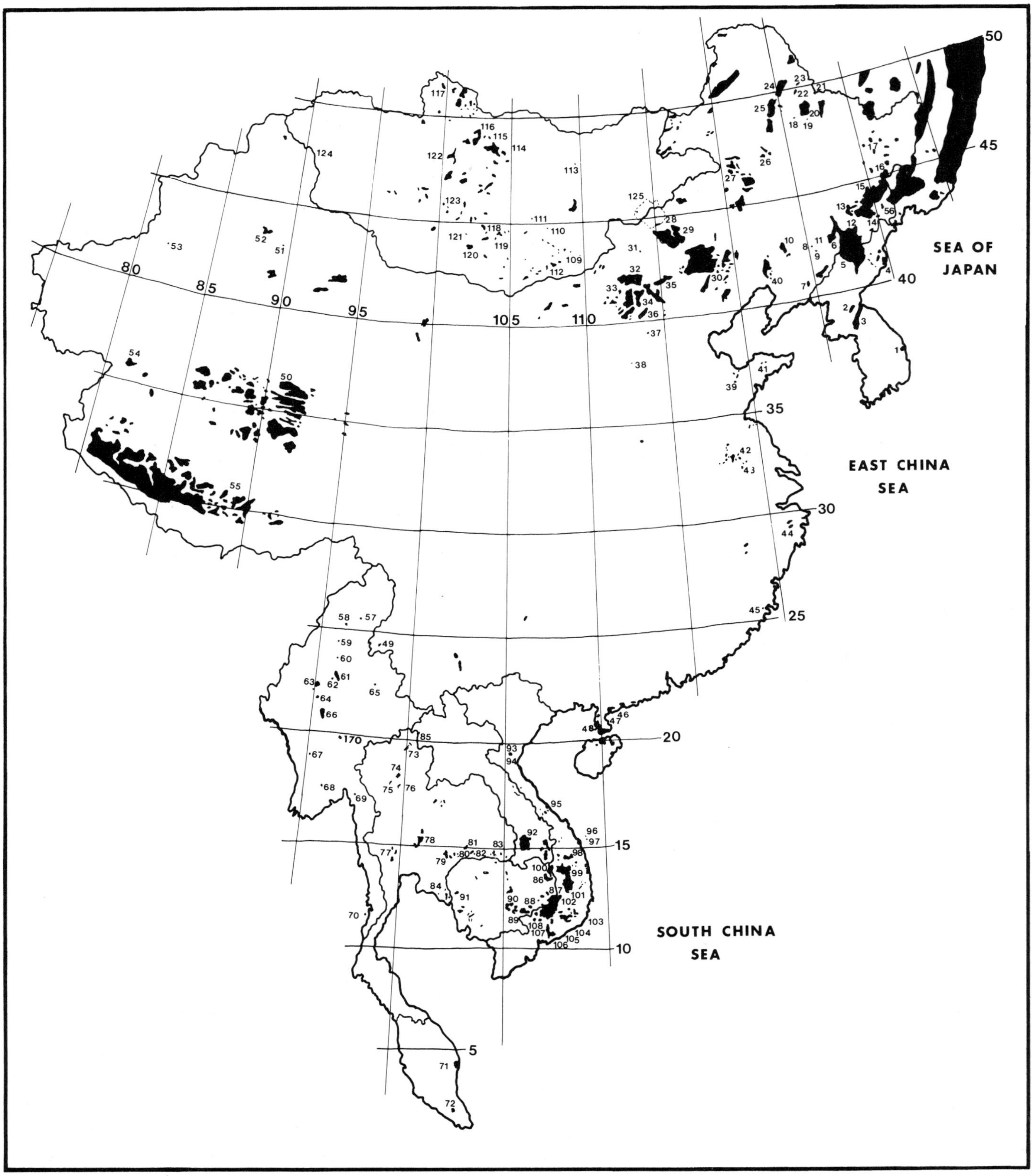

Figure 1a. Distribution of exposed Cenozoic volcanic rocks in southern mainland Asia shown in black. Numbers correspond to the localities described in Part 1 of this paper. Note that there are a large number of areas for which no information has been found.

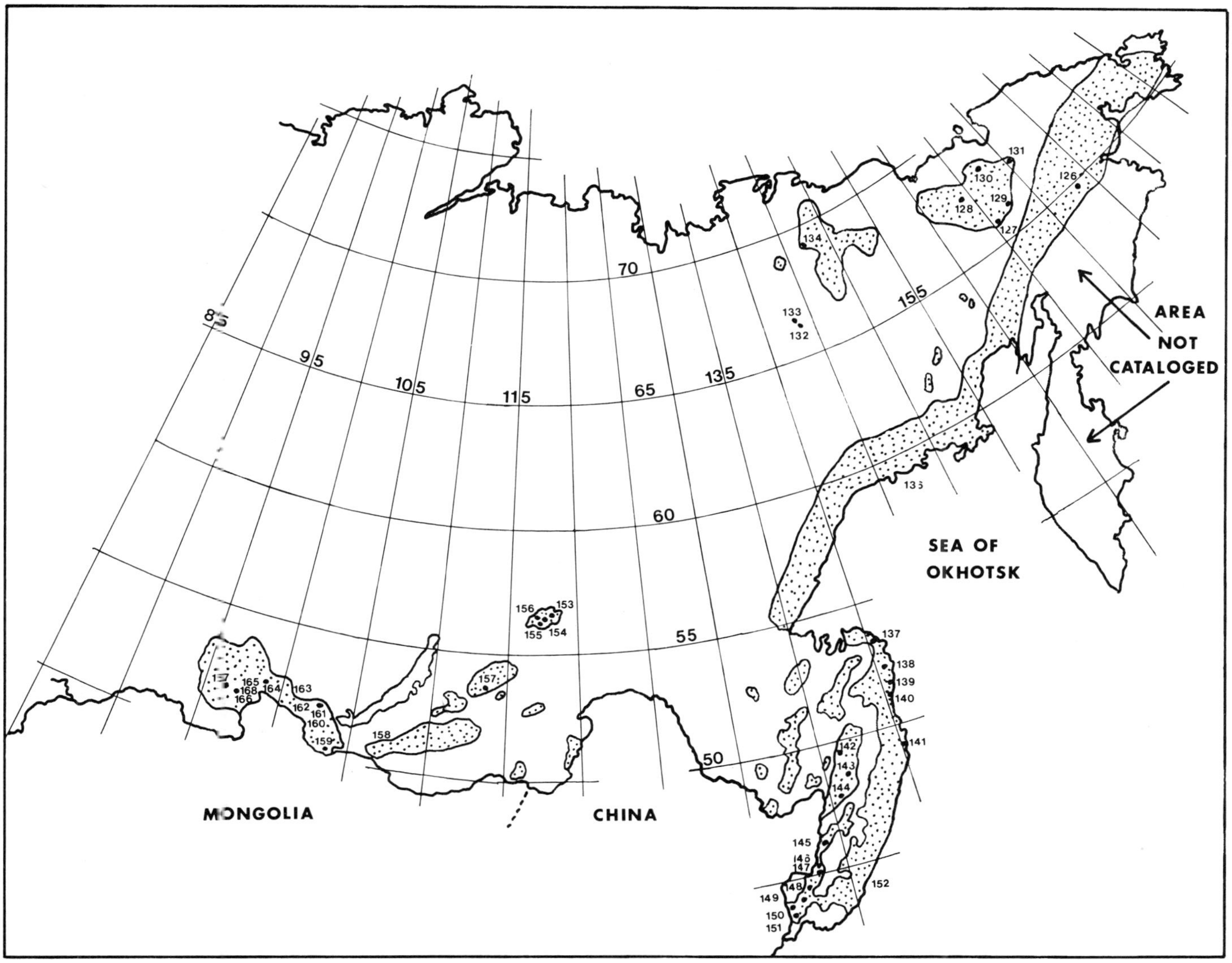

Figure 1b. Paleogeographic distribution of Cenozoic volcanic rocks in the eastern Soviet Union. (Data for both figures compiled from: Barr and Macdonald (1981), Chhibber (1934), Devyatkin and Smelov (1979), Grenard (1929), IAVCEI (1973), Kuno (1953), Li (1970), Ogura (1969), Saurin (1967), Sengor and Kidd (1979), Stephenson and Marshall (1984), Tateiwa (1960), Tokyo Geological Society (1929), UNESCO (1972), Vinogradov (1967).

bracketed. There are at least five English representations of Chinese characters. The most commonly employed in the past was the Wade-Giles system. The present work gives locality names in the modern Pin Yin system. Translation from the former to the present system is critically dependent on the location of apostrophes; poor typesetting has often resulting in the published name bearing little resemblance to the real name. Where possible, I have translated directly from the original Chinese characters to the Pin Yin system. This has not been possible for all the localities; those are therefore appended with question marks.

PART 1

PRODUCTS OF CENOZOIC VOLCANISM IN ASIA

KOREA

During Cretaceous time, the landscape of Korea was first subjected to violent eruptions of andesite, and later of liparite. These were associated with a compressional tectonic regime. The style of tectonomagmatic activity changed during the Cenozoic, the tectonics becoming tensional and the erupted products basaltic (Kobayashi, 1956). During the Cenozoic there were four main centers of activity in Korea. These areas are widely separated, in three regions, ano do not appear to be tectonically related.

*1. **Yongil area.*** This was a center of eruptive activity during the Miocene period in which the Ŏil basalts were extruded. These are largely olivine basalts, although a two-pyroxene andesite was also produced (Tomita, 1969).

*2. **Koksan-Singye.*** According to Tomita (1969), this area contains one of the largest lava flows in eastern Asia, being some 45 km from north to south and 20 km from east to west. The flow parallels the Ch'uga-ryŏng flow to the east, running in a northeast-southwest direction. The lava is of late Pleistocene age and may be associated with a shield volcano.

*3. **Ch'uga-ryŏng.*** The extensive lava flows of this area are enclosed by a graben. Tomita (1969) described one of the flows as being 60 km in length and apparently erupted from an unnamed Late Pleistocene shield volcano. Another flow is 40 km in length and was derived from a shield volcano named Ap-san. Ap-san is approximately 100 m in height, has a summit crater 200 m wide and 20 m deep, and has more than 20 parasitic cones developed on it. Recent work (Lee, 1981) to the south of the DMZ (Demilitarized Zone) has led to the recognition of between 2 and 11 separate flows that are Late Pleistocene to early Holocene in age (Fig. 2). The lavas all appear to be alkali olivine basalts.

*4. **Kilchu-Myŏngch'ŏn.*** In this area the lavas varied to a greater extent in terms of their chemical composition and were erupted during a longer period of time. A sequence of analcite basalts and olivine trachybasalts were erupted in early Miocene time. The olivine basalts are extremely rich in potassium. The largest of these flows supposedly extends more than 50 km and is about 1,000 m thick! Alkaline trachytes and comendites were erupted during the Pliocene period. The alkaline trachytes continued to be erupted during the Early Pleistocene, but the eruptions changed to quartz-normative basalts during the Late Pleistocene (Tomita, 1969).

CHINA-KOREA BORDER

*5. **P'aektu-san (Baitoushan).*** (Also spelled as Baitoushan, Baegdu, and Baekdoosan, and known in Japan as Hakuto.) This is a 2,751-m-high strato-volcano resting on a shield and basaltic plateau. The shield on which the 700-m-high cone rests is known as the Chang-pai-shan (Changbaishan) and Lao-hei-tingtzu (Laoheidingzi) volcano. The earliest structure formed was the basalt plateau, which is approximately 150 km in diameter and rises from approximately 1,000 to 2,000 m above sea level. The basalt is an augite-olivine variety. The alkaline trachyte and pantelleritic trachytes that form the upper part of the aspite (=shield) were erupted during the Pliocene. The trachyte is 30 to 40 km in diameter and up to 300 m thick, the lavas sloping away from the center of the volcano with a dip of 10°. During the Late Pleistocene, long flows of pantellerites, tholeiites, quartz basalts, and olivine-basalts produced the upper dome of the volcano. This has a basal diameter of 10 km and is more than 400 m thick with slopes of 20 to 30° and individual flows ranging from 30 to 40 to more than 300 m in thickness. The next stage in the development of the volcano was the production of a number of parasitic cones. Kim (1979) plotted the distribution of these cones from LANDSAT imagery and showed that they tend to be aligned N75°E ± 5°. Ogura (1969) recorded the two cinder cones Dayanzhifeng (Ta-yen-chih-feng) and Xiaoyanzhifeng (Hsiao-yen-chih-feng) to the southeast of P'aektu and noted that there are three cones averaging 30 m in height and 1 km in basal width on the north side of the volcano summit crater and several other cones to the west.

One of the most recent events on P'aektu was the eruption of a pumice that has a radius of 40 km centered on the summit caldera. Blocks of trachybasalt and nordmarkite were ejected with the pumice. It is probable that this eruption preceded and initiated the formation of the summit caldera. The caldera is psuedo-octagonal, measures 5 km from east to west, and has a depth of 869 m from the caldera rim to the lake floor. The crater lake (Ch'ŏn-ji, Tien-chi, Tianchi) measures 3 km from east to west, 4 km from north to south, covers an area of 9.2 km^2, and is 375 m deep. There are unsubstantiated reports that P'aektu erupted in 1597 and 1702 (Ogura, 1969).

Tomita (1969) also recorded several mountains composed of volcanic material rising to the south-southeast or southeast of P'aektu. These include Hwang-bong and Kwandu-bong, both composed of alkaline rhyolite, and Soboek-san, Pukp'ot'ae-san, Changgun-bong, P'aeksa-bong, Namsŏl-lyŏng, and Turyusan, all composed of alkaline trachyte. Also see Plate 317 in Short and others (1976). P'aektu is located approximately 42°N, 128°E.

CHINA

Cenozoic volcanics crop out in a large number of areas in China, particularly in the northeast, and in Tibet. As some of these areas have suffered considerable erosion since the deposition of the volcanics, in a number of cases it is difficult to ascertain the original dimensions of the erupted products. There is also a great variety in the styles, compositions, and volumes of the eruptions. Because of their remoteness and inaccessibility, some areas have only recently been examined on the ground or have been mapped only from LANDSAT images.

6. Lung-wan (Longwan, Longgang, Chingyu) Group. The Holocene-aged Longwan group lies 130 km to the west-northwest of P'aektu and comprises an area of olivine-basalt and trachybasalt lavas some 50 km in diameter. These lavas were derived from more than 35 cinder cones (Ogura, 1969: see Appendix 6 for measurements of some cones and Fig. 3). The cinder cones are fairly typical, with basal diameters of up to 1.5 km and heights of up to 79 m. Ogura (1969) mentioned flows from four of the craters. There are two flows from the Dayishan cone that extend 6 km downvalley and are 2.5 km wide. The upper flow is 7 m thick; the lower is 13 m thick. Ogura (1969) also recorded a thick flow to the east of Dayishan that is covered by a layer of lapilli. The lavas occupy an area of 11.5 km^2. The flow from Dagushanzi also occupies an area of 11.5 km^2, and is 9 km long and 1 km wide. Additional lava was added to this flow from cones situated to the immediate northeast. The third and largest flow erupted near the Jinlongdingzi cone and flowed west-southwest, 35 km along the floor of a river valley. The flow varies in width from 200 m to 5 km, is 20 m thick in its lower course, has an area of 66 km^2, and has a blocky surface. The fourth flow, which is 8 km long and 17 m thick, is considerably eroded; its source was not identified. Ogura also recorded several unsurveyed flows to the north of this volcano group. A group of 20 cinder cones was mentioned by Ogura (1969) as being near the Longwan Group, but he provided no more information about them. As many as 72 cones may be in the Longwan area (Dr. Zhu to J. H. Latter, personal communication, 1983), located around 42°N, 126°E.

7. Huang-i-shan or K'uan-tien (Huangyishan, Kuandian). According to Ogura (1969), there are several cones at this locality; he mentions three. Huangyishan is a cone 513 m above sea level and has a relative height of 100 m (IAVCEI, 1973). The cone is truncated and composed of alkali olivine basalt. Lu and others (1981) stated that the volcanic rocks are up to 91 m thick and can be divided into four eruptive cycles. Cycles I and IV were mainly lavas, whereas cycles II and III began with intense pyroclastic explosions and terminated with lava flows. From the base upward the lavas consist of basanites, olivine alkali basalts, and olivine tholeiites. Abundant inclusions of lherzolite, harzburgite, and pyroxenite, and megacrysts of garnet, clinopyroxene, and anorthoclase are found. Jingyishan is also a truncated cone, and there is a cone farther to the north of Jingyishan. Analyses by Cong Bolin and others (1979) and Zhou and Armstrong (1982) appear to be from this locality and include strongly alkaline trachybasalt and alkaline hawaiite. The area is located at approximately 40.75°N, 124.6°E.

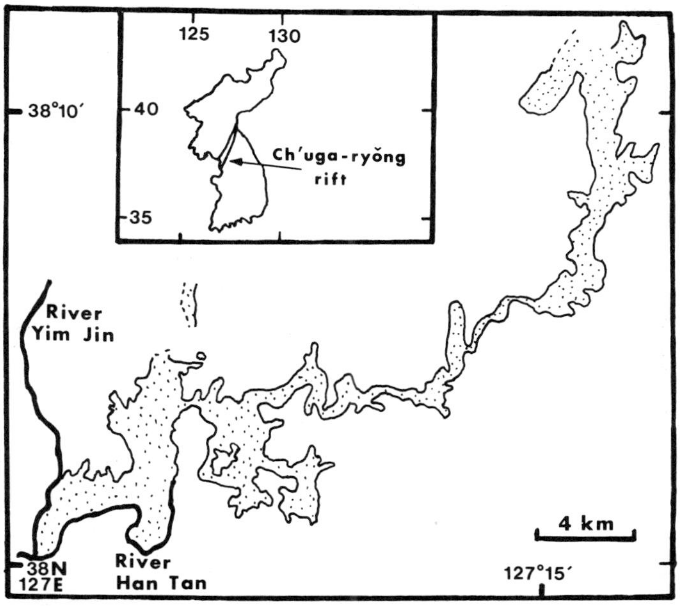

Figure 2. Outline of basalt southwest of Ch'uga-ryong rift (listing 3), Korea. (Modified after Lee [1981].)

8. Ta-t'un or Fan-chia-tung (Datun, Fanjiatun). Ogura (1969) identified four domes or shield volcanoes, which are all less than 40 m in height: Jianshen, Pingdingshan, Datundongshan, and Dadongbinglingshan. They are composed of olivine basalt, are arranged in two parallel lines trending northeast-southwest, and are located approximately 43.5°N, 125°E.

9. I-t'ung (Yitong). Six basaltic domes arranged in two parallel lines running northeast-southwest are located in this area. On the one line are the volcanoes Dongjianshan, Xijianshan, Dagushan, and Xiaogushan, and on the other are Maanshan and Molijingshan. These domes are about 50 m in height (Ogura, 1969) and are located at approximately 43.25°N, 125°E.

10. Ch'i-hsing-shan (Qixingshan). There are seven domes of Holocene age in this area. They have heights of less than 110 m and diameters of less than 900 m (see Appendix 6 for dimensions and Fig. 4). Most are composed of olivine basalt but Naobaoshan is composed of leucite-basanite. They are located about 43.9°N, 123.9°E.

11. Ying-o-men (Yingemen). Tsung-pu (1960) mentioned the early discovery of the nepheline basalt of Ts'ao-shih (Caoshi) at this locality. The location from his map is uncertain, although it could be within the Longwan area.

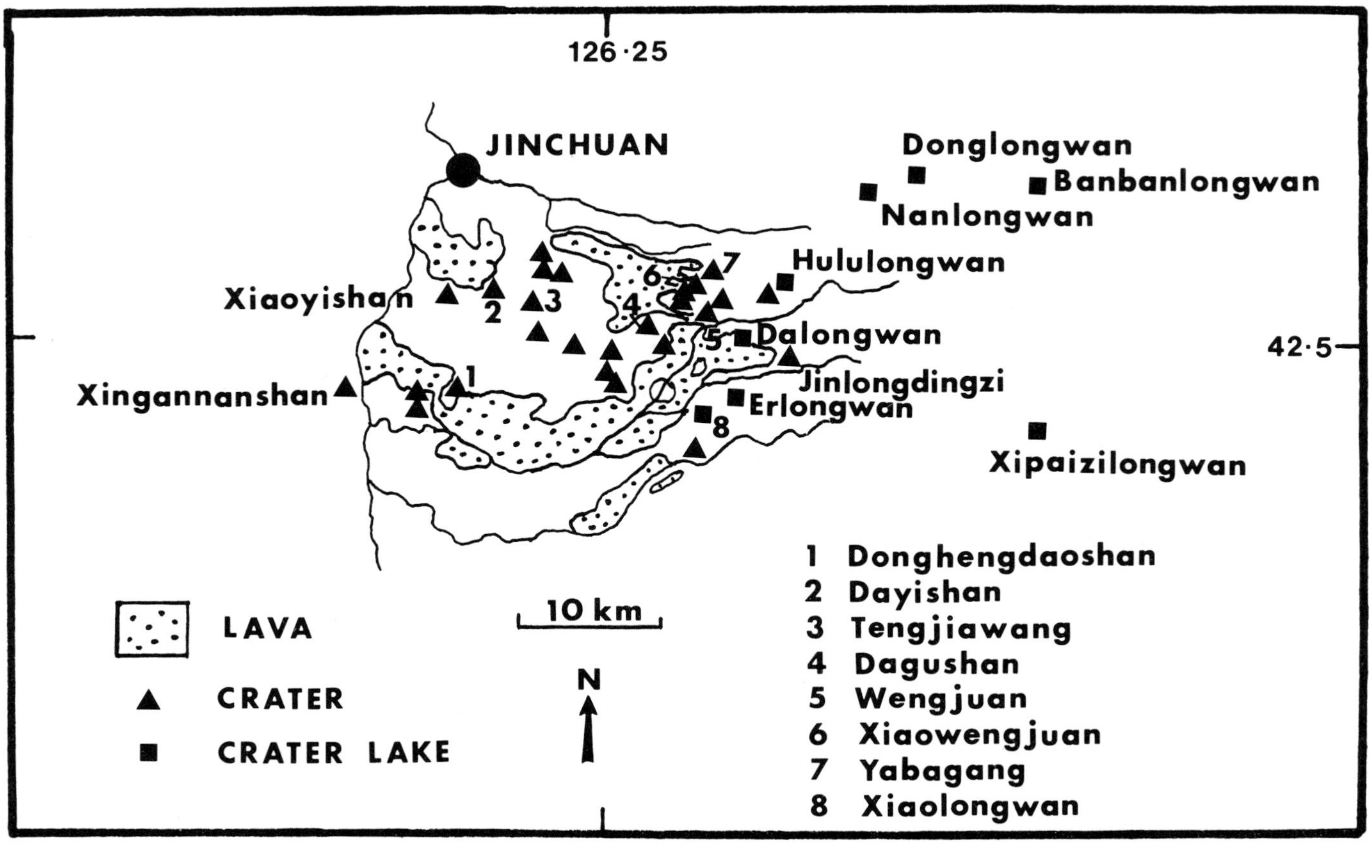

Figure 3. Distribution of volcanic features in Longwan district (listing 6) of Manchuria. Modified after Ogura [1969].)

12. Tun-hua (Dunhua). This little-known area is approximately 40 km in diameter, is composed of flows of two or three different ages (Kuno, 1953), and is centered at approximately 43.25°N, 128°E.

13. Ching-pe (Chingpohu; Jingbo). To the northwest of the lake there is a flow of olivine-rich alkali basalt 100 km long that flowed along a canyon. The flow has an average thickness of 10 m and a width of 5 to 10 km (Kuno, 1953). Shen and Wang (1983) stated that volcanism in this area can be divided into four separate episodes; Pliocene, Middle Pleistocene, Late Pleistocene, and Holocene. Holocene basalts have been dated at 5,410 years in age by the ^{14}C method and 8,300 years by the thermoluminescence method. These authors noted that various tablelands along the Mudanjiang river are composed of basalt and that there are many Holocene cones on the tablelands. The lavas exhibit such features as dammed lakes, falls, and lava tubes. The lavas include olivine latite basalt, phonolitic basanite, and tephritic phonolite. Ultramafic inclusions are common near the cones and include dunite, spinel lherzolite, and saxonite. The distribution of volcanism has been controlled by the northeast-striking Dunha-Mishan fault. The location is approximately 44°N, 129°E.

14. Ningan to Yenchi. This is a severely dissected plateau of Middle to Upper Tertiary age. The lavas are preserved as widely separated mesas and buttes of olivine basalt. Lavas extend over an area of approximately 120 km trending north-south and 60 km trending east-west centered on 43.5°N, 129.5°E (Kuno, 1953). E Molan and others (1981) described inclusion-rich basanites from this general area. The inclusions are lherzolite and olivine clinopyroxenite.

15. Ta-ping-shan (Dabingshan). This has been described as a large shield volcano with an east-west diameter of 60 km and a north-south diameter of 90 km. It has a height of 1,000 m and is composed of alkali-olivine basalt, probably of middle to upper Cenozoic age. There are a few minor flat cones (shields?) northeast of the higher part of the shield (Kuno, 1953). The location is centered at approximately 44.25°N, 130°E.

16. Unnamed (45 N, 130.5 E). A lava plateau extending approximately 100 km north-northeast–south-southwest and 40 km east-west, again probably of middle to upper Cenozoic age. The rock is a fine-grained, alkali-olivine basalt.

17. Sung-hua-chiang (Songhuajiang). A series of four basalt domes located along the eastern side of the river, one of which reaches a height of 100 m. One of these domes was called

Makuli (Maguli?) by Ogura (1969). The location is approximately 47°N, 131°E.

18. Chien-shan (Jianshan). This is a small cone to the west of the Wudalianchi Group (20), about 20 m in height and composed of basaltic materials (IAVCEI, 1973). Its location is 48.5°N, 125.6°E.

19. Erh-ko-shan (Erkeshan). This consists of a 33-km^2 leucite-basanite plateau or low shield (slopes of 1 to 2°) with three truncated cinder cones of Holocene age. The largest cone is 110 m high and has a basal diameter of 300 m (see Appendix 6 for measurements). The cones are the same composition as the underlying plateau (Ogura, 1969). It is located at approximately 48°N, 126.3°E.

20. Wu-ta-lien-ch'ih (Wudalianchi). Also incorrectly referred to as Uyunhordongi, this area had a fairly complex history that Ogura (1969) separated into five stages: (1) early plateau lava, (2) shield-like plateau lava, (3) older cone lavas, (4) Shih-lung (Shilong) lava, and (5) young cone lava. According to Ogura (1969), the plateau lavas cover an area of about 500 km^2 and are composed of anorthoclase-olivine trachybasalt that was erupted from several centers. Tomita (1969) dated this series of flows as being of Holocene age. The overlying shield volcanoes have a maximum thickness of 130 m and slopes of 1 to 2°. The shield lavas are the same composition as the plateau basalts, differing only in their lack of macroscopic olivine. The 12 old cones have heights of about 100 m and basal diameters of 1 km. Ogura (1969) argued that they were arranged in an orthogonal network (Fig. 5). The lavas are leucite basalts. The Shilong lava is largely an obsidian pahoehoe flow with lava tubes. It covers an area of 68 km^2, and has a maximum width of about 10 km and a maximum length of about 18 km. The outline of the flow is like that of a tennis raquet. The lava is a leucite basanite. The two new cones, Laoheishan and Huoshaoshan, erupted between 1719 and 1721. They have dimensions similar to those of the early cones and are composed of similar material.

Recent studies (Feng and others, 1979; Feng and Whitford-Stark, 1986) have also indicated five periods of eruption: Lower, Middle, and Upper Pleistocene; Holocene; and historic. There is no evidence for a plateau basalt preceding the development of the 14 cones. About 85 percent of the Shilong lava was derived from the Laoheishan cone, the remainder from Huoshaoshan. The Laoheishan lavas average about 16.5 m in thickness, whereas those of Huoshaoshan are about 8.5 m thick. The vents for the lava were fissures that cut through the two cones. The lava exhibits a variety of pahoehoe surface features and also has tumuli, hornitos, and lava falls all indicating emplacement as a low-viscosity material. The location is centered at approximately 48.6°N, 126.25°E. Ogura (1969) also mentioned the Nomoer volcano, consisting of two cones along the river to the east of Tetu.

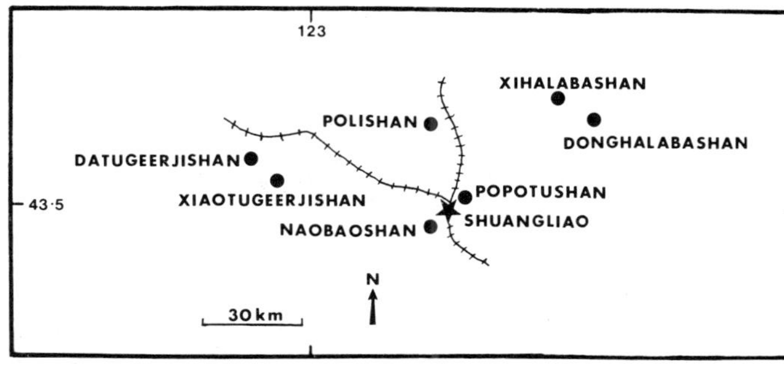

Figure 4. Distribution of volcanic features (filled circles) in Qixingshan district (listing 10) of Manchuria. Shuangliao is a city on the railroad line (Modified after Ogura [1969].)

21. Ih-chia-shan (Yijiashan ?). The IAVCEI (1973) catalog lists this as a lava cone 100 m high and composed of basalt. Located at 49.6°N, 127.6°E, it may be the volcano Nomoer (see listing 20, above).

22. K'o-lo-nan-shan (Keluonanshan). The Keluo group consists of a series of five cones arranged in a north-east-southwest line, the largest of which is 200 m in height (see Appendix 6). Another cone, Xishan, is located 30 km northeast of Keluo, and there is an unnamed cone to the south-southwest on the upper course of the Zaolaihe (Ogura, 1969). These cones are basaltic in composition. The location is centered on 49.25°N, 126°E.

23. Chia-ts'sung or Ka-tsung (Jiacong or Gacong). There are three cones, again arranged along a northeast-southwest direction, two of which are 150 m in height (Ogura, 1969). These cones are basaltic in composition. The location is approximately 49.8°N, 126.8°E.

24. Kan-k'uei (Gankui or Ganhe). Ogura recorded the presence of four or five cones with relative heights of 170 m. There are possibly as many as seven, composed of rizzonite (a variety of limburgite) and olivine basalt (Dr. Zhu to J. H. Latter, personal communication, 1983). The location is approximately 49.6 N, 124.6 E.

25. Sha-t'u or Sa-ti-erh (Shatu or Sadier or Shaheershan). This is a series of eight cones scattered over a distance of 170 km north-northwest–south-southeast along the Nominho (Nuominhe). These cones range in height from 100 to 450 m and are composed of olivine basalt (see Appendix 6) (Ogura, 1969). The location is approximately 49.2°N, 123.6°E.

26. Onenor. This is an explosion crater now filled by a lake 2 km in diameter with surrounding heights of 250 to 300 m. The walls of the crater are composed of liparite (Ogura, 1969). The location is approximately 47.5°N, 121.25°E.

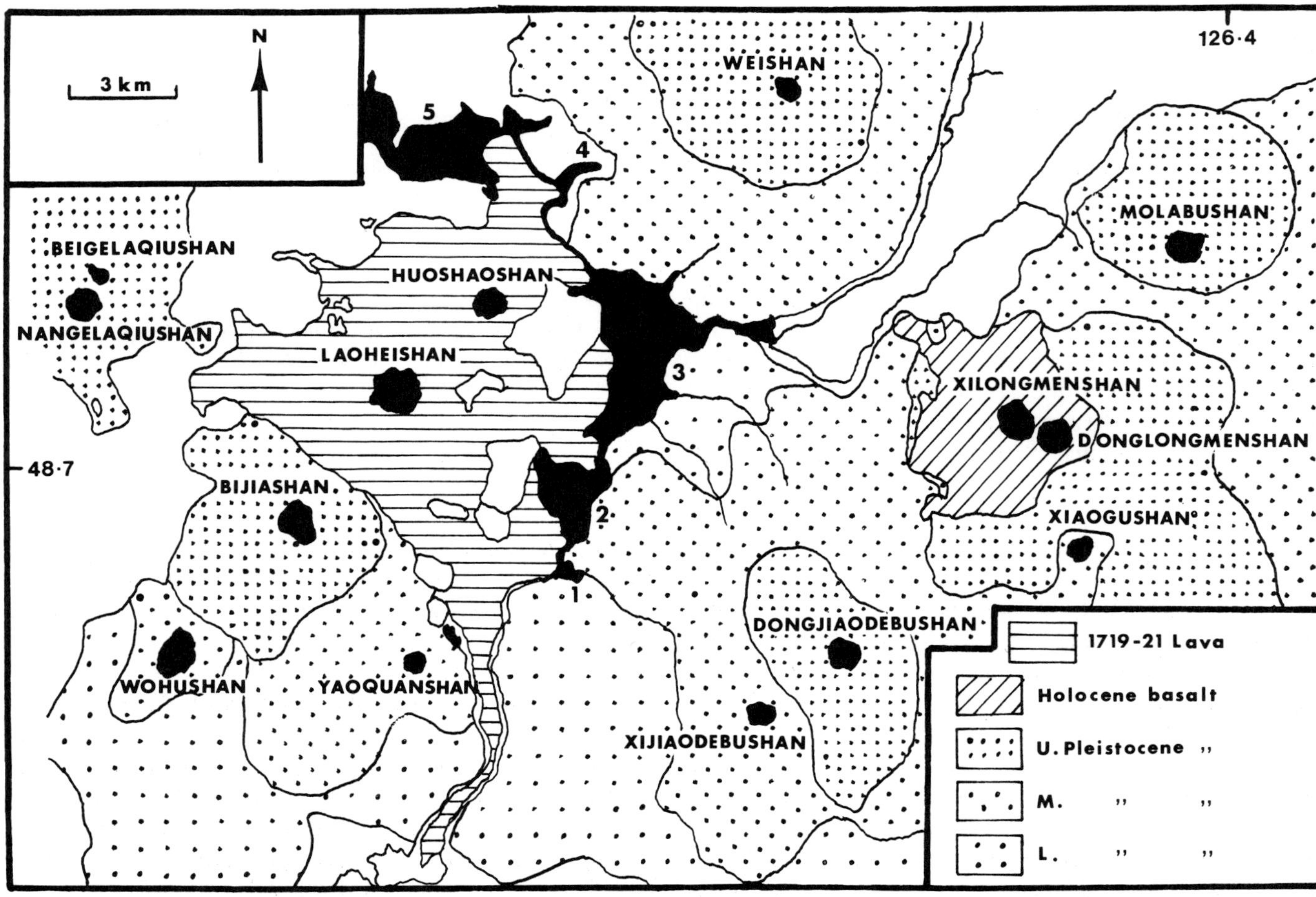

Figure 5. Distribution of volcanic features in Wudalianchi (five connected lakes) district (listing 20) of Heilongjiang Province, northeastern China. Lakes are shown in black and are numbered; other features shaded in black are cinder cones. The 1719–1721 lava is known as the Shilong flow. (Reproduced from Feng Maoseng and Whitford-Stark [1986]; reproduced with permission of Elsevier).

27. Ha-r-hin-tun (Baerqian ?). This volcano is probably of early Pleistocene age and is 1 km in diameter and 200 m high. The summit crater contains two cones, and the volcano is composed of quartz-trachyte, liparite tuff, and obsidian (Ogura, 1969). The location is very approximately 48°N, 119.8°E.

28,29. Dolon Nor (Dalibo). The earliest description of this area by Teilhard de Chardin (1926) called it Dalinor. He was not, however, describing the area around Lake Dalai-nor (now Hulunchi) near the border of the Soviet Union, but rather the area around the lake, which in most modern atlases is called Ta-er-hu or Tarinur, near the border of Mongolia. The IAVCEI (1973) catalog listed about 20 to 30 cinder cones ranging in height from 70 to 150 m in this area. A map compiled from LANDSAT imagery shows that the cones are arranged in northeast-southwest alignments, as was also noted by Ogura (1969). Teilhard de Chardin (1926) wrote that he observed in excess of 40 fresh volcanoes with both single and double cones and separation distances of 3 to 4 km. Teilhard de Chardin did not see any preferential alignment but described the volcanoes as being randomly distributed over an area 60 by 25 km. He also noted that the rocks he sampled from localities up to 45 km distant were all black, glassy limburgites, with augite and olivine microlathes set in a brown glass. In addition, he dated the volcanoes of Quaternary age and stated that there was a maximum of two episodes of magma eruption. Tapponnier and Molnar (1977) noted that individual flows in this area (which they called Hsing-an; see Fig. 6) were huge (perhaps as much as 20,000 km^2 in area), and their fresh appearance suggested that they may even have erupted within historic time. The location is approximately 43.5°N, 116.25°E (see also listing 125 below).

30. Wei-chang (Weizhang). The Weizhang basalts are of Upper Miocene age and are up to 75 m thick (Lee, 1939). They are part of the Hanruoba basalt (see listing 35). The location is approximately 41.95°N, 117.6°E.

31. Unnamed (43.6°N, 112.6°E). The IAVCEI (1973)

catalog lists this area as containing nine basaltic cones on a lava field.

32, 33. Unnamed (45.75°N, 112.6°–113°E). This location consists of two groups of Pleistocene-aged cinder cones; one group contains two cones and the other three (IAVCEI, 1973).

34. Honggeertu (E. Yinshan [Tsining]) (41.5°N, 113°E). This is a group of about 12 cones of Holocene age, 7 of which are aligned along a fault (IAVCEI, 1973). According to Dr Zhu (personal communication to J. H. Latter), there are six volcanoes. Tapponnier and Molnar (1977) showed a series of flows associated with cinder cones in this general area.

35. Hanoorpa (Han-jo-pa; Hanruoba). As described by Barbour (1929), this marks the southern limits of a dissected lava plateau (Kalgan area). Remnant mesas of basalt-capped hills stand out in front of the main scarp. The Hanruoba basalt is some 150 m thick and is probably of Late Oligocene age. The lava is an olivine basalt, sometimes with trachyte porphyry, containing olivine nodules. The location is approximately 40.9°N, 115°E.

36. Ta-t'ung (Datong). This location consists of a group of 10 small cones ranging in height from 60 to 100 m and containing craters. They are dated as end Middle Pleistocene in age and are composed of a black, low-viscosity olivine basalt (Tomita, 1969). According to Dr. Zhu (personal communication to J. H Latter), there are 12 cones, including Pailoushan, Jiujinshansi, Shuangshan, and Bujishan. *Chinese Earthquakes* (1981) recorded the presence of 20 cones, of which 13 are "well formed" and named the volcanoes Xizijulebao, Jinshan, Heishan, Langwoshan, Henwoshan, and Eshougedashan. The publication also reported hot gas emerging from one of the Datong volcanoes. The location is approximately 40.1°N, 113.2°E.

37. Unnamed. The IAVCEI (1973) catalog lists a volcano at 39°15′N and 113°31′E. Tapponnier and Molnar (1977) showed two flows at 40°15′N and 113°31′E, which may be the same area. Cong Bolin and others (1979) gave an analysis of a Quaternary basalt from the Jingxing-Juanbei area.

38. Central Taihang Range (Daihang). This area consists of numerous small patches of basalt containing decomposed olivine and occurring as units 20 to 30 m in thickness. The flows are probably of Oligocene age (Lee, 1939). Zhou and Armstrong (1982) gave an analysis of an alkali basalt from 36°N, 114.2°E.

39. Tai-an, Ssu-hui, Fei-hsien (Taian, Sihui, Feixian). Scattered outcrops of a post-Oligocene basalt are found in these three counties (Lee, 1939). Zhou and Armstrong (1982) presented an analysis of an alkali basalt from 36.5°N, 118.5°E (see Appendix 7).

40. Fu-shun (Fuxin). This is a series of upper Eocene

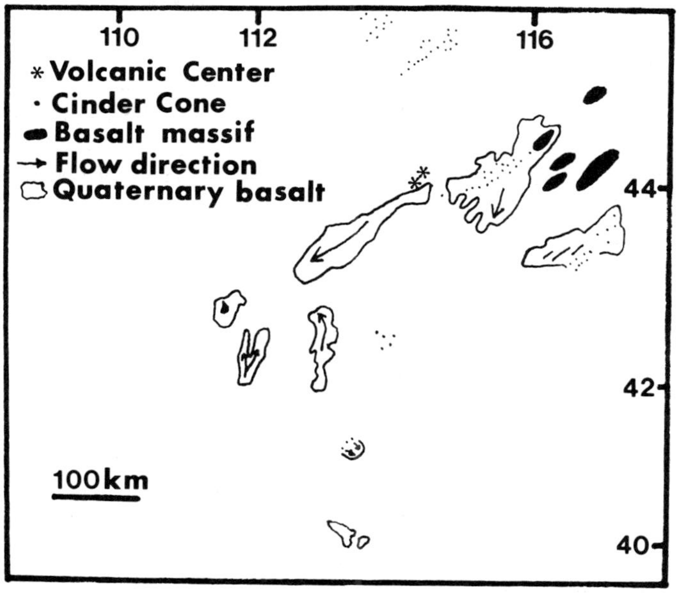

Figure 6. LANDSAT interpretation of volcanic features of Hsing-an (Xingan) district (listing 28) of Manchuria and Mongolia. (Modified after Tapponnier and Molnar [1977].)

basaltic flows that have a groundmass very rich in potassium. The rock was described as a black porphyry called "kokuhangan" by Tomita (1969). Cong Bolin and others (1979) presented an analysis of a late Tertiary basalt from near Jingyuan and early Tertiary basalts from along the Xialiaohe. Zhou and Armstrong (1982) presented analyses of subalkalic basalt and picrite from approximately 41°N, 122°E.

41. Ch'i-hsia, T'ai-shan (Qixia, Taishan). Tsung pu (1960) described a nepheline basalt existing at these two localities. Likewise, Zhou and Armstrong (1982) presented analyses of strongly alkaline nephelinites from approximately 37.5°N, 120.85°E (see Appendix 7).

42. Nushan. This locality is described in the IAVCEI (1973) catalog as a truncated basaltic cone with a height of 50 to 60 m and a Pliocene to Lower Pleistocene age. The location is 33.1°N, 118.75°E.

43. Fangshan. Anderson (1923) described a number of hills to the northeast of Nanjing as being of basaltic composition. Originally these were interpreted to be eroded volcanoes, but Anderson reinterpreted them as being isolated basalt-capped mesas from a once-extensive lava sheet. He conceded that one of the hills (Fangshan) could represent an eroded cinder cone. The hill is 75 to 100 m in height and of Lower Pleistocene age (IAVCEI, 1973). The rock is an olivine basalt with a potassium-rich andesine (Tsung-pu, 1960). It is located at approximately 31.95°N, 118.95°E.

44. Eastern Chekiang (Zhejiang). Lee (1939) described

a 30-m-thick basalt rich in olivine and magnetite from a locality at approximately 29.5°N, 121°E.

45. Haicheng (Haizheng). Lee (1939) described a 200-m-thick unit consisting of a lower volcanic agglomerate, basalt flows, and an upper tuffaceous sandstone sequence of possibly lower Tertiary age at approximately 25°N, 119°E. Zhou and Armstrong (1982) presented an analysis of a Quaternary subalkalic picrite from 23.3°N, 116.15°E.

46. Shandun (Leizhou Bandao, Luichow Peninsula). This is a group of nine Holocene-aged tholeiitic volcanoes described as destructive (IAVCEI, 1973). The named volcanoes include Mazhang, Qixingling, Nanxing, Lingshi, Shiwanshan, and Jiyunling, located from 20.5° to 21°N, 109° to 110.5°E.

47. Khaynanj (Hainan Dao). This is a group of seven tholeiitic cones up to 250 m in height and of Lower to Upper Pleistocene age (IAVCEI, 1973), which are located from 19.5° to 19.95°N, 109.5° to 110.75°E. According to Dr. Zhu (personal communication to J. H. Latter), there are eight cones, including Gaoshanling, Leihufeng, Yiongxing, Huangzo, and Longtang. Zhou and Armstrong (1982) presented an analysis of an alkaline andesite from 20.3°N, 110.15°E.

48. Wei-chou-tao (Weizhoudao). Situated in the Gulf of Tonkin, this is a circular, shield-like island 4 to 5 km in diameter with a 1-km-wide explosion crater in its southern part. The rock is an anorthoclase-bearing, olivine trachybasalt of Pleistocene(?) age (Tomite, 1969). The location is 21°N, 109°E.

49. Teng-yueh (Tengchong). The volcanic rocks of this district were divided into three distinct groups: an early series of bedded andesites, a sequence of massive andesites, and a late sequence of olivine basalts (Burton, 1913). On the basis of their "freshness," the andesites were thought to be of Tertiary age. The field relations of the bedded andesites are complex, and they have been folded by the Tertiary orogeny that affected the area (Coggin brown, 1913). The older andesites are pyroxene-bearing whereas the massive andesites are commonly amphibole-bearing.

The olivine basalts were interpreted to be of Late Pliocene to Pleistocene age and to have issued from four eruptive centers (see Fig. 7). Maanshan (called She-toe-shan, Shejiushan in modern translation, by Coggin Brown and also known by the name of Hawshuenshan) is a cone composed of ropy and scoriaceous basalt that rises 275 m above the level of the surrounding plain. The upper 92 m of the volcano has a steeper slope, and the summit crater wall is missing on the northern side. The volcano occupies an elongate area approximately 9.4 by 6.4 km in extent (Coggin Brown, 1913). Laoguibo is older than Maanshan, rises 244 m above the plain, and has an almost circular summit crater 30 m deep and 182 m in diameter. The west side of the crater wall is missing. Dayingshan occupies an area of about 7.5 by 7.5 km, and rises 2,865 m above sea level and 990 m above the surrounding plain. The southeast wall of the crater is missing. The volcano is composed of black, slaggy pumice-like lavas.

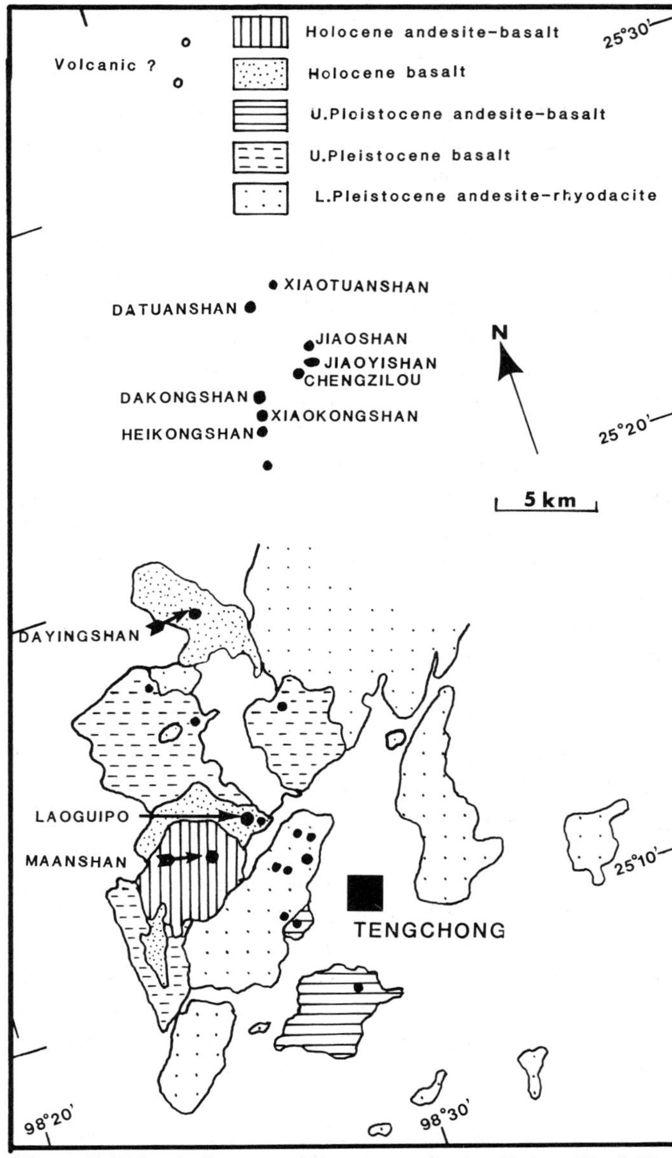

Figure 7. Volcanic features of Tengchong district (listing 49) of West Yunnan, China, compiled from LANDSAT imagery. Volcanic units in southern part of area simplified after Wei Junchao and others (1980). Names for volcanic vents, outlined in black, supplied by Feng Maoseng. Two conical features in the north may be volcanic.

The Kung-po (Gungpo) volcanoes were described (Coggin Brown, 1913) as a series of seven small puy-like volcanoes up to 250 m in height associated with lavas covering an area of 5.6 by 3.2 km to the north of Tengchong. The southernmost of the cones are heavily vegetated, but the northernmost appear to be much younger and have very little vegetation and individual flows are clearly visible. Coggin Brown also noted that hills to the north of the area had the appearance of volcanoes: Yaowushan (2,194 m above sea level), which may be composed of andesite, and Luzongshan (2,682 m above sea level). *China Pictorial* (1980) contained a photo of the Dalong crater lake and mentioned that about 60 cones are in the Tengchong district, 20 of them intact. Yang Qilong and Shen Minzi (1981) stated that the Tengchong cones were produced in five cycles from the Pliocene to the Quaternary and that the area is a site of active geothermal fields. The five periods of volcanism identified by Wei Junchao and others (1980) and Zhu and others (1983) are: (1) Early Pliocene (4.0 to 7.2 m.y.) flood eruptions of pyroxene basalt, (2) Late Pliocene (3.4 m.y.) eruptions of olivine basalt, (3) Early Pleistocene (0.98 m.y.) olivine basalts and andesite-dacites, (4) Late Pleistocene (0.15 to 0.9 m.y.) andesitic basalts, and (5) an Early Holocene stage of pyroxene olivine basalt and a Late Holocene (0 to 0.05 m.y.) stage of andesitic basalts. *Chinese Earthquakes* (1981) reported that the travel notes of Xuxia from May 1639 recorded what appears to have been geyser activity, and that in 1609 some 500 or 600 sheep and several shepherds were struck by lightning. Furthermore, the aborigines had told him that Dayingshan mountain had at that time been covered by huge trees and bamboos, all of which were burned in fires that continued for several days. Additionally, deep pools on the mountain had dried up. The Yong-Chang annals also recorded continuous earthquake activity through January and February of 1609. This appears to have been a volcanic eruption. Other large earthquakes of 1512 and 1343 were not apparently associated with any volcanic activity. The location is around 25°N, 98.5°E.

50. Northern Tibet. Tapponnier and Molnar (1977) depicted three conspicuous calderas near the western end of the Kunlun fault based on LANDSAT imagery. Kidd (1973), also using LANDSAT imagery, identified an eruption column at 35°51′N, 91°42′E coming from a crater within a 1.5-km-diameter cone situated on a ring fracture of a 4-km-diameter caldera. The crater is at an altitude of 5,200 m. The terrain appears to be composed of potassium-rich andesites, dacites, and rhyolites. Maps of the U.S. Defense Mapping Agency denote the volcano Debussey in this general area. Several extinct volcanoes in this general region are also depicted on the map by Grenard (1929 p. 339) that is based on the expeditions of early explorers.

Recently the area around 35.5°N, 87°E has been the location of field work by Chinese geologists. Deng Wanming (1978a) described three areas—Bamaoqiongzong, Yongbohu, and Qiangbaqian. All three are characterized as having Hawaiian-style volcanism of Quaternary age. Both table mountains and conical constructs are present.

Bamaoqiongzong, the southernmost of the three areas, extends more than 20 km north-south and 15 km east-west, covering an area of 300 km^2, with the main summit area rising 5,400 m above sea level. There are several isolated flows separated from the main body, some of which form flat-topped, columnar jointed areas. One lava structure is more than 100 m high. It was estimated that the inner area originally rose 300 m higher than it is now. A perfectly preserved edifice is located to the northeast of the main summit. It rises 150 m above the surroundings and is conical in form with outer slopes of 22°. The crater is concave in form and has been breached on the south side by lava flows. The crater is 230 m in diameter, more than 20 m deep, has a slope of 17°, and is 5 m wide at the bottom where there is a pit. Another edifice lies to the southeast of the main summit. A lava flow 5 m thick overlies Quaternary lake deposits of fine sands and gravels. The volcanic rocks are peralkaline and include noselite phonolite, pseudoleucite porphyry, and analcitite.

Yongbohu consists of five separate volcanic bodies. The main body has a relative height of 150 m. Another flat-topped body is elongate, like a dragon, in a southeastward direction. Another is cone-shaped. The oldest rocks are sandstones, overlain by Tertiary red beds, overlain by Quaternary lavas. The rocks are augite dacites, trachyandesites, and augite andesites.

Qiangbaqian stretches east-west along the southern border of the Kunlun mountain range. It is about 55 to 70 km wide from north to south. Quaternary volcanic rocks stretch from the southern border of the Kunlun to Yongbocuo. The most important are black, glassy smooth lava flows. Both tube-fed and fissure flows are present; the latter are widespread. In the north, a volcano with flanks not more than 200 m high has a columnar jointed top. Color photographs of these volcanics can be found in the work of Deng Wanming (1978b).

51. Turfan (Hotcheou). Humboldt (1830) described this as an isolated volcanic cone that emits a column of fire. It is located approximately 42.9°N, 89.25°E. It is written in the History of the Song Dynasty that, "a smog often sprang up from the mountain, but there are not clouds and fog. The blaze looked like the blaze of a torch at night. The skin of animals turned red when the blaze lit them." (*Chinese Earthquakes,* 1981).

52. Ou-roumtsi (Urumchi). This is described by Humboldt (1830) as a solfatoric area with a radius of 5.3 km. Also near Urumchi (30 li to the west of Byrke boulak) is an area that Humboldt called the "flaming plain." (Note: The text uses both the French word *lieues* for league, which has variable dimensions, and the Chinese word *li,* which is approximately 0.5 km. Humboldt set 200 li as equal to 15 lieues; the two occasionally appear to have been confused.) He also described an area that he called the "hole of cinders," 45 km in circumference, on the border of the Ili Province and the Urumchi district. The Illi volcano was recorded in the *Book of Wei* as follows, "There is a volcano in the south border of Illi. The stones on the side of the volcano were scorched melt. It coagulated after it had flowed several miles. People make use of it for medicine—so-called sulphur stone" (*Chinese Earthquakes,* 1981). Bonney (1899) expressed doubts as to the volcanic origin of some of these features, noting that the Katou solfatara was, in reality, ignited coal beds. Urumchi is located at 43.8°N, 87.75°E.

53. Pechan (Peishan, Baishan). (Also known as Hochan, Aghie, Bichbalick, Khala, and Boschan.) This was described by Humboldt (1830) as a volcano, whose name in

English means white mountain, near the town of Kouche (41.7°N, 82.6°E). Humboldt noted that the name of the mountain was based on the color of the rock rather than on the presence of snow. He also stated that the mountain has extensive lava fields and is recorded as having erupted in the first and seventh centuries.

54. ***Vulcan.*** Located at 36°N, 81°E, this is a volcanic cone that rises some 6,000 m above sea level and is probably composed of basalts and latites. A number of craters and vast lava fields are known to exist in the area. Vulcan is recorded as having erupted in 1951 (Vlodavetz and Piip, 1959). Deng Wanming (1978b) published a photo of a conical volcano in the Kardas volcanic district, which he stated erupted in 1951. Some Chinese geologists consider the 1951 eruption doubtful because no new products of magmatic origin were found (Feng Maoseng, personal communication, 1981). Also, Norin (1979) described Hill 5891 (35.5°N, 82.95°E) as a dome-shaped volcano with latitic lavas of monzonitic magma type.

55. ***Southern Tibet.*** Sengor and Kidd (1979) have identified two extensive zones of Cenozoic volcanism within Tibet from LANDSAT imagery. The southern zone (Fig. 1) comprises an area some 1,000 km east-west and 400 km north-south, of which it is estimated volcanic rocks form about 60 percent of the terrain. These volcanics could date back to Upper Cretaceous in age. (The northern zone is described in part in listing 50.) Also identified from LANDSAT have been ignimbrite plateaus and calderas. Rocks sampled from Tibet include andesites, dacites, and rhyolites, and more recently, alkali basalts.

56. ***South of Tungning.*** This area constitutes a large lava plateau (about 75 by 120 km) straddling the Chinese-Soviet border at 44°N, 131°E. The plateau rises from about 200 to 800 m above sea level and has a slope of about 1°. As Tertiary gravels underlie the plateau, it is considered to be of mid- to late Tertiary age. The plateau is estimated to be about 150 m thick near the center and is composed of several flow units (7 and 15 of which have been recorded from different localities), which are usually less than 10 m thick although they can range up to 20 m thick. The lavas are basalts, tholeiites, and quartz dolerites. Near the summit of the plateau are shield volcanoes; these are composed of olivine and pyroxene basalt, are about 1 km in diameter and less than 100 m high, and are without summit craters (Kuno, 1953).

BURMA

The Cenozoic rocks of Burma lie largely along parallel north-south trends through the center of the country. The principal tectonic influences have been the subduction of the oceanic Indian plate and post-Miocene dextral movement along the Sagaing fault subsequent to the initiation of development of the Andaman basin (Stephenson and Marshall, 1984). The volcanic features take a variety of forms and, for the most part are predominantly more andesitic than those previously described (except Tibet).

57. ***Mount Loiyme (Mount Loimje).*** This is a breccia and lava dome that was active during the Tertiary period. The mountain rises 1,561 m above sea level, of which about the upper 1,000 m are composed of volcanic materials. Ashes and tuffs are mixed with basalt, augite andesite, and doleritic lavas (Chhibber, 1934). It is located approximately 26°N, 96.5°E.

58. ***Jade Mines.*** The volcanic rocks at this locality are Pliocene to Pleistocene in age. There is a great variety in composition, including an early peridotite and serpentinite plateau, Hill 1361 of olivine basalt, and a 4-km-long exposure of horizontal siliceous agglomerate. The magma changed in composition from a picrite-basalt to rhyolite, both with time and with distance westward from the eastern border (Chhibber, 1934). The extrusives were probably erupted through fissures. It is located at approximately 25.75°N, 96.5°E.

59. ***Taungthalon (Taungthonlon).*** A volcanic mountain rising 1,706 m above sea level, about which little is known (Chhibber, 1934). The area consists of basaltic and andesitic lavas, tuffs, ashes, and agglomerates (Bender, 1983). Stephenson and Marshall (1984) presented an analysis of a hypersthene-augite-hornblende andesite from the mountain. The location is 24.95°N, 95.75°E.

60. ***Wuntho.*** This is a series of domes, or conical hills, or piles of lava with craters. The rocks are largely tuffs and ashes with dips of 30°, which were possibly emplaced subaqueously, with occasional basalt flows. Three domes are named by Chhibber (1934): Shinmetaung, an elongate breccia dome; Minletaung, a horseshoe-shaped dome; and Pinotaung. These were dated as Tertiary in age. Bender (1983) also reported possible Lower Oligocene andesites and trachytes in the southeast portion of the district and basalt dikes near Pinlebu of possible Oligo-Miocene age. The location is 21.2°N, 96.5°E.

61. ***Singu Plateau.*** This Recent lava plateau occupies an area of 62 km^2, about 18.8 by 4.8 km, and is composed of mugearite and teschenite alkaline extrusives. The lavas were probably erupted from a fissure (Chhibber, 1934). The location is 22.5°N, 96.1°E.

62. ***Shwebo.*** A series of Pliocene-age, decomposed basalts interbedded with rocks of the Irrawaddian Series (Chhibber, 1934). The Irrawaddy Group has since been found to be Middle Miocene to Lower Quaternary in age (Bender, 1983). A 4.2-km-length of basaltic lava can be traced in the Kabwet district. This lava has a pillow structure (Chhibber, 1934). The location is 22.6°N, 95.6°E.

63. ***Lower Chindwin.*** This area contains a series of seven

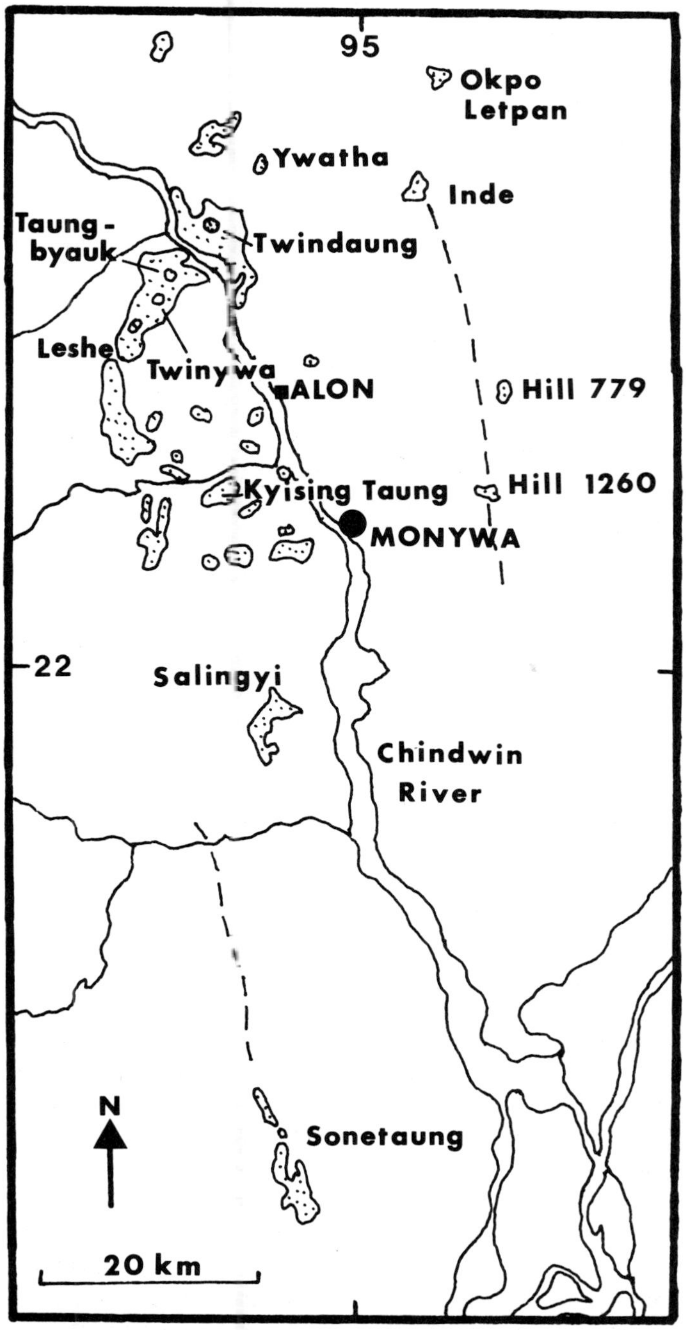

Figure 8. Distribution of volcanic materials in Lower Chindwin area (listing 63) of Burma. Dotted areas are where volcanics crop out. Capital letters indicate names of towns. (After Chhibber [1934] and Bender [1983].)

or eight explosion craters of Late Pliocene and Pleistocene age (Fig. 8). Ywatha is a circular depression some 850 m in diameter and 45 m deep with a flat floor. Basaltic blocks and ashes are found on the walls and floor. Twindaung stands some 150 m above the level of the river. Its crater has dimensions of 1.45 by 1.1 km, and is some 121 m from rim to water level, which is 22 m deep in the crater. The edifice is composed largely of ash and debris with large xenoliths and blocks of olivine basalt and hornblende andesite. Xenoliths include peridotite, pyroxenite, and perknite. Taungbyauk is some 800 m in diameter and 60 m in depth, and is composed mainly of ashes and blocks of olivine basalt (this appears to be the same feature that Bender [1983] labeled Silaung Taung). South of this is a poorly defined crater and south of that is the Twinywa crater. Twinywa is 1.2 km in diameter and 60.9 m in depth. The ash beds are 10 to 20 m thick and contain blocks of alkali olivine basalt. Leshe craters are compound with a superficial cover of ash deposits.

Apart from the explosion craters, there are scattered outcrops of basaltic lavas in the district. At Natlabo there is a lava sheet 3 to 6 m thick and 0.8 by 0.6 km in extent. South of Minma are two hills called Myetthadaung, which are capped by a sheet of olivine basalt. At Silaung Taung there is a plateau about 2.0 by 1.2 km in size, capped by a 15-m-thick olivine basalt with no apparent vents. West of Silaung Taung is another plateau capped by a highly decomposed basalt. At Songyaung (opposite Alon) there is an extinct crater now occupied by a lake. The rocks in this area are olivine basalts, lapilli ash, and tuff. A very recent basalt forms a plateau to the north of Thayet-pingan, 2.5 by 1.6 km in area, possibly originally fed from a central vent. Rhyolites and quartz porphyries form the Kyaukmyet Hills, Myayeik Taung, Hills 937, 943, 994, 995, and 1055, and outcrops west of the Kabauk village. Dacites, dolerites, and diorites form the Salingyi uplands. Andesites are found near the ninth mile on the Yinmabin road (Chhibber, 1934). Bender (1983) reported a K/Ar age of 5.8 m.y. for a hornblende porphyry from Kyising Taung. The area is located around 22.1°N, 94.8°E.

64. East Monywa. At Kyaukkadaung there is a basaltic core to a hill that measures 1.6 by 0.4 km. Picrite basalt also forms the core of Hill 779 and Thuzi Hill. Inde Hill appears to be the site of an explosion crater composed of ash with blocks of olivine basalt. Okpo Letpan is a hill 120 m high with a capping of decomposed olivine basalt. Natyindaung is an isolated mass forming two hills comprising a dike and an extrusion of olivine basalt (Chhibber, 1934). The location is approximately 22.1°N, 95.1°E.

65. Loi Han Hun. This is a conical hill rising about 213 m above the surrounding plain. The upper part of the hill is formed of a dome of olivine basalt of Late Pliocene to Pleistocene age (Chhibber, 1934). Bender (1983) indicates that it is probably a columnar jointed plug. It is located at approximately 22.5°N, 98.2°E.

66. Mount Popa. According to Chhibber (1934), the Mount Popa complex occupies an area of 1,280 km^2 (35 km north-south, 20 km east-west) and consists of a cone with a summit crater that has been removed on the northwest side and a number of subsidiary vents. The volcano rises 1,518 m above sea-level and 914 m above its base (Fig. 9). The summit crater is 30 m deep and appears to be about 1.6 by 0.8 km in size. Activity

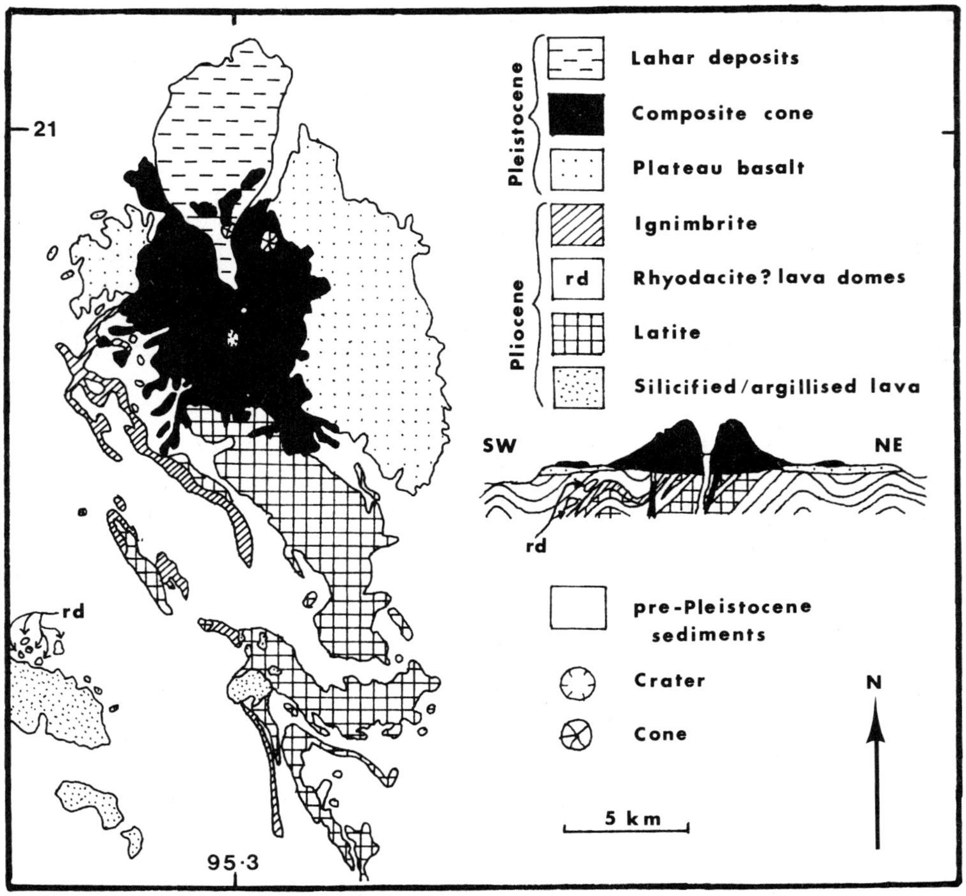

Figure 9. Outline map and schematic cross section of Mount Popa region (listing 66) of Burma. (Modified after Stephenson and Marshall [1984].)

apparently began in the Pliocene with an eruption of andesite onto a plateau. The rocks forming the Mount Popa cone are largely post-Pliocene and are hornblende andesites and olivine basalts. Vesuvianite was recorded.

The flows average about 3 m in thickness and the last eruption of the volcano was probably in 442 BC (after Chhibber, 1934). A detailed survey of the volcano by Stephenson and Marshall (1984) separated the extrusives into Pliocene-aged Older Volcanics and Pleistocene-aged Younger Volcanics. The Older Volcanics cover an area of 73 km^2 to the south of Mount Popa. The lavas are microporphyritic latites with some flows having paleosols. Pyroclastics are rare in this sequence and no obvious source has been identified. Overlying the highest flows and resting on a paleosol is a widespread, nonwelded ignimbrite. The ignimbrite crops out over an area of 22 by 5 km and could constitute the product of a single eruption. In places the ignimbrite has a basal agglomerate with blocks of rhyodacite up to 1.25 m in diameter. Coarse rhyodacite also forms nearby conical hills that may represent lava domes. The Younger Volcanics form Mount Popa, which Stephenson and Marshall (1984) described as rising some 1,150 m above the surrounding plain and consisting of a 25 to 30° composite cone with an area of 57 km^2 surrounded by a 69-km^2 lava plateau. The summit crater is 1.6 km wide and 850 m deep (note the discrepancy between these values and the earlier values of Chhibber). The breach on the northern side led to the production of a 3-km^3 lahar. The cone consists of alternating basalt and basaltic andesite flows, coarse pyroclastic deposits, and scoria. The majority of the flows are porphyritic augite-basalt, with lesser amounts of augite-amphibole basaltic andesite and a few flows of amphibole-basaltic andesite. The majority of the eruptions appeared to be of vulcanian type with strombolian activity dominating the later stages of development. The surrounding plateau consists of blocky flows of coarsely porphyritic augite-basalt, often with a layer of pyroclastic material at the base of each flow. The volcano is located at 21°N, 95.2°E.

67. Prome. There are a series of four olivine dolerite intrusions of post-Miocene age in this area (Chhibber, 1934). Also known as the Pegu Yoma sills (Stephenson and Marshall, 1984). The location is around 18.6°N, 95.1°E.

68. Tharrawaddy. A series of five intrusions of serpentinized olivine dolerite of post-Miocene age were recorded by

Chhibber (1934). Basaltic volcanics have recently been encountered in boreholes in this area (Stephenson and Marshall, 1984). The location is around 17.75°N, 95.8°E.

*69. **Thaton.*** There is a band of rhyolites and rhyolitic tuffs, about 30 to 800 m in width and about 9.6 km long, that has been dated to be of Late Pliocene to Pleistocene age (Chhibber, 1934). The location is around 16.9°N, 97.4°E.

*70. **Medaw Island.*** There is a plateau 80 m high composed of olivine basalt and occupying the west side of the island. The basalt is probably of late Tertiary to sub-Recent age (Chhibber, 1934). It is located at 11.7°N, 98.6°E. Other islands in the Andaman Sea have not been included here.

*170. **Yezin, Pyinmana.*** Small basalt outcrops have been found at this locality—19.75°N, 96.25°E (Stephenson and Marshall, 1984).

MALAY PENINSULA

Located some 450 km northeast of the currently operative subduction zone that has produced the Sunda Arc, the Malay Peninsula has witnessed very little Cenozoic volcanism.

*71. **Kuantan.*** A series of basalt flows crop out over an area of 150 km^2. The basalts are less than 136 m thick and may have been fissure-fed. The rocks are nepheline basanites and olivine basalts (Gobbett and Hutchison, 1973). A K-Ar age of 1.6 m.y. has been derived for the basalts (Barr and Macdonald, 1978). The location is around 3.9°N, 103.4°E.

*72. **Segamat.*** There are three flows: a leucite tephrite, a leucite basanite, and a potassic ankaramite (Grubb, 1965). The flows vary from 18 to 42 m in thickness, and the upper meter of each flow is laterized. The lavas occupy an area of approximately 50 km^2 (Gobbett and Hutchison, 1973). Originally believed to be contemporaneous with the Kuantan lavas, recent K-Ar dating has yielded an age of at least 62 m.y. for the Segamat flows (Barr and Macdonald, 1981). The location is around 2.5°N, 102.8°E.

THAILAND

Both basaltic and rhyolitic volcanics are found in Thailand. Until recently the basalts were believed to be of Tertiary age (Brown and others, 1951), and the andesitic and rhyolitic rocks were ascribed a similar or younger age. Jacobson and others (1969), however, believed that the andesites and rhyolites were of early Triassic age. A recent analysis of the granites within Thailand (Beckinsale and others, 1979) indicated granitic magmatism of Lower Triassic, Upper Triassic, Lower Cretaceous, and Middle Cretaceous age. Thus it would appear that all of the Cenozoic volcanism in Thailand has been of basaltic composition. A location map for the Indochina volcanic deposits is presented in Figure 10.

*73. **Ban Chiang Khian.*** One of the few tholeiites found in outcrop within Thailand, this area consists of dissected flows located within hilly terrain (Barr and Macdonald, 1978). It has an age of 1.62 ± 1.25 m.y. (Barr and Macdonald, 1981). The location is approximately 19.8°N, 100°E.

*74. **Lampang.*** This is one of the largest basalt areas within Thailand, extending about 20 km along the railway line and covering about 200 km^2. The area is flat lying on a valley floor and overlies gravel deposits containing pebble tools made by early humans (Barr and others, 1976). At one locality, an upper section of basalt 3 m thick is underlain by subaqueous foreset pillows probably resulting from flow into a lake. Farther west the basalt is a massive columnar flow. From one to seven flows are exposed at other localities. The flows are believed to have been derived from one or more vents located in the northeast part of the field (Barr and others, 1976). Paleomagnetic studies show the early flows to have a reverse polarity and the younger flows normal polarity. This, plus inconclusive fission track and K-Ar dating, led Barr and others (1976) to propose that the younger basanites are about 0.64 m.y. old, and the older, 0.69 to 0.89 m.y. in age. Alternatively, the younger flows could be 0.89 to 0.95 m.y. in age, and the older flows in excess of 0.95 m.y. Barr and Macdonald (1981) listed the Palad volcanic vent near the Lampang to Mae Moh highway. The location is around 18.3°N, 99.6°E.

*75. **Sop Prap.*** This consists of dissected flows of hawaiite located in hilly terrain around 17.9°N, 99.5°E (Barr and Macdonald, 1978).

*76. **Denchai.*** A series of flows covering 70 km^2 occurs to the southwest of Denchai. The flows follow a northeast-trending lineament and are about 17 km long and 4 km wide, with the vent area inferred to be close to the center. Seven separate flows with a combined maximum thickness of 25 m have been identified. The earliest, most extensive flow was an alkali olivine basalt that has a maximum thickness of 5 m. This event was followed by two flows of hawaiite, each 2 to 3 m thick. Two later hawaiites, with a combined thickness of 8 m, contain a 1-m-thick horizon of ultramafic nodules (largely spinel lherzolites). The final unit is a basanite 3 to 4 m thick, with abundant nodules up to 10 cm in diameter. This flow appears to be the source for gem-quality sapphires and zircons. All the flows have reversed polarity and the youngest has an age of 5.64 ± 0.28 m.y. A paleosol that developed between the youngest and oldest flows may indicate that not all flows were erupted within the same reversal interval (Barr and Macdonald, 1979). The location is around 18°N, 100°E. Brown and others (1951) also reported an outcrop of basalt to the southwest of the town of Sukhothai (17°N, 99.6°E).

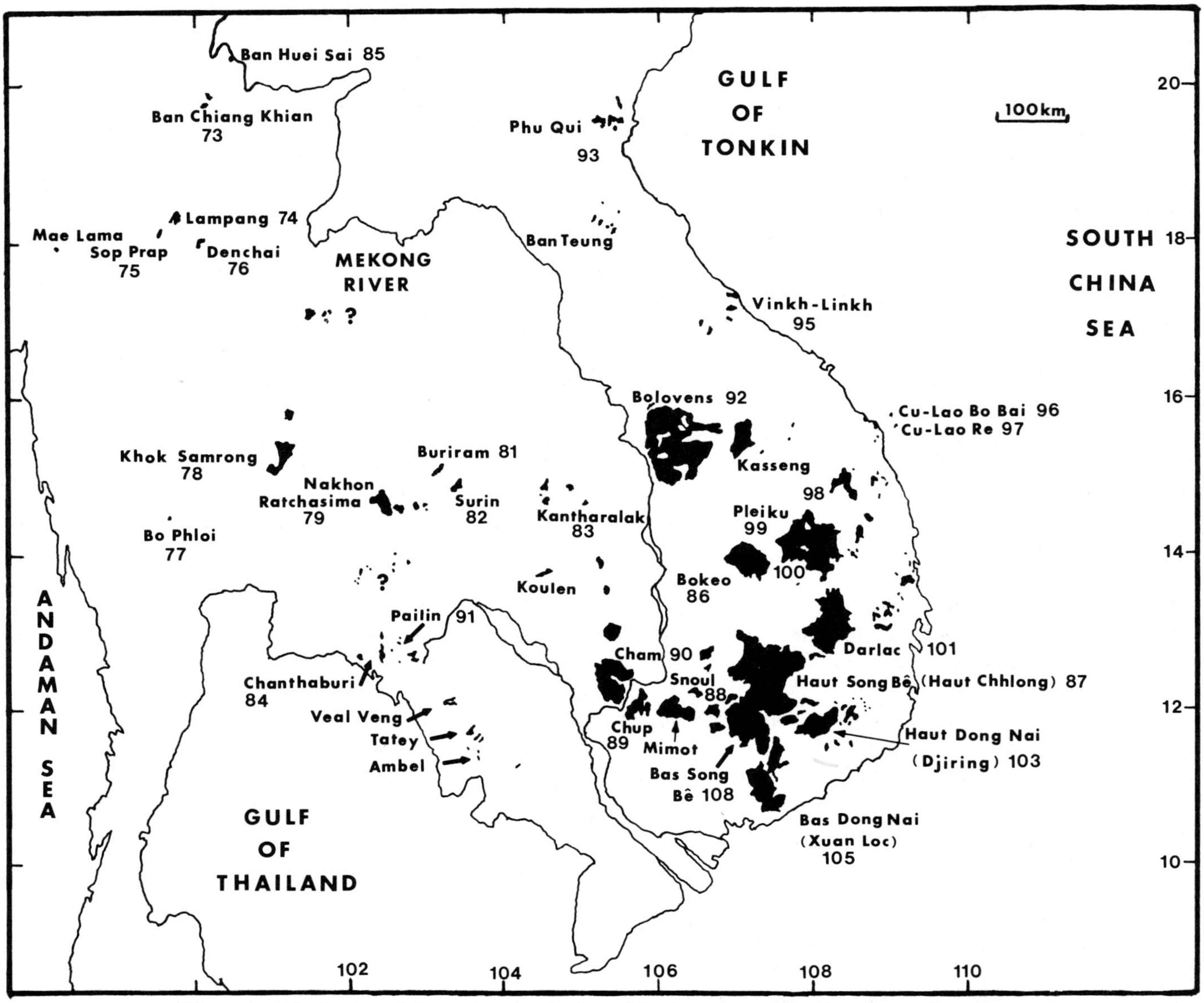

Figure 10. Distribution of Cenozoic volcanics in mainland southeast Asia. Question marks refer to localities where volcanics are doubted. Name of nearest town has been given to unnamed localities. (Compiled from Barr and Macdonald [1981], IAVCEI [1973], and Saurin [1967].)

77. Bo Phloi. A nepheline hawaiite, rich in spinel-lherzolite nodules, megacrysts, and many inclusions, forms a semicircular plug 300 m in diameter. Gravels surrounding the plug contain gem-quality sapphires and spinels (Barr and Macdonald, 1978). A K-Ar age of 3.14 ± 0.17 m.y. has been obtained for the plug (Barr and Macdonald, 1981). Brown and others (1951) recorded the existence of a basalt outcrop to the northwest of the town of Lop Buri, as well as scattered plugs and flows of nepheline olivine basalt near Kanchanaburi. The location is around 14.5°N, 100°E.

78. Kok Sam Rong. Flows of alkali olivine basalt occur in association with hawaiites (Barr and Macdonald, 1978). The basalts occupy valleys in hilly terrain. An age of 11.29 ± 0.64 m.y. has been obtained for a tholeiite (Barr and Macdonald, 1981). The location is 15.2°N, 100.7°E.

79. Nakhon Ratchasima. Sheet-like flows of hawaiite occur in this area, but they are covered by a red soil several meters thick and are poorly exposed (Barr and Macdonald, 1978). The location is at 15°N, 102.1°E.

80. Pradonchai. La Moreaux and others (1958) reported the presence of a flow 10 by 5 km to the southwest of Pradonchai. Barr and Macdonald (1978) found that the flows in this area were hawaiites, in places overlain by a mugearite. The location is at 14.7°N, 103°E.

81. Buriram. A basalt flow some 20 by 60 km in extent is situated 5 km to the southwest of the town (La Moreaux and others, 1958). The basalt is a hawaiite derived from a central vent (Barr and Macdonald, 1978) and has a K-Ar age of 0.92 ± 0.30 m.y. (Barr and Macdonald, 1981). The location is at approximately 15°N, 103°E.

82. Surin. A basalt flow that becomes progressively thicker from north to south occupies an area of 16 by 5 km (La Moreaux and others, 1958). The flow, a mugearite, was derived from a central vent eruption. The location is around 14.9°N, 103.4°E.

83. Kantharalak. La Moreaux and others (1958) reported two basaltic flows near Nam Om; one with dimensions of 20 by 4 km, located to the northwest of the town, and the other with dimensions of 15 by 4 km, located to the northeast. One of these would appear to be the nepheline hawaiite described by Barr and Macdonald (1978). Farther south, gem-quality corundum and zircon are found associated with olivine basalt cobbles, but no outcrops were observed (Barr and Macdonald, 1978). To the west are a sheet flow of hawaiite and a nepheline mugearite plug 500 m in diameter. The hawaiite, which is poorly exposed, is covered by a thick, red soil. The plug has a K-Ar age of 3.28 ± 0.48 m.y. and is the only nepheline mugeearite in Thailand (Barr and Macdonald, 1981). It is older than, and not associated with, the nearby hawaiite (Barr and Macdonald, 1978). The location is around 14.6°N, 104.6°E.

84. Chanthaburi Region. There is a nephelinite flow from a central vent volcano in the Tha Mai area to the west of Chanthaburi (Barr and Macdonald, 1978). The Khao Phloi Waen cone at Tha Mai has a K-Ar age of 0.44 ± 0.11 m.y. (Barr and Macdonald, 1981). Zircons, presumably from the flow (Kra), give anomalously old fission track ages of 29 ± 1.7 and 23.6 ± 1.2 m.y (Carbonnel and others, 1972). To the east, sapphires and rubies are mined at the western and southern margins of lobate, north-trending nephelinite lava fields. Flows to the north of the field are transitional to basanite (Barr and Macdonald, 1978). A zircon from one of the basalts to the east of Chanthaburi gave a fission track age of 2.57 ± 0.2 m.y. (Carbonnel and others, 1972). The location is around 12.5°N, 102.2°E.

85. Ban Hui Sai. Alkali olivine basalts form dike-like bodies in the Mekong River near the town, and a K-Ar age of 1.74 ± 0.12 m.y. has been obtained for a subcircular outcrop of basanite 50 m in diameter at Chiang Kong on the Thai side of the border (Barr and Macdonald, 1981). The location is at 20.3°N, 100.3°E.

LAOS AND KAMPUCHEA (CAMBODIA):

There are extensive flows of Quaternary-age olivine and nepheline basalts in these two countries. Most of the geologic studies in the region were made by the French, largely in the early part of this century.

86. Bokeo (Ratanakiri). This basalt plateau was the subject of a detailed study by Lacombe (1969). The plateau is the fifth largest in Indochina with an altitude varying from 100 to 500 m. Currently it covers an area of 1,500 km^2 and has a volume of 90 to 100 km^3. Lacombe estimated that the basalt may have covered 3,200 km^2 and had a volume of 150 km^3 prior to erosion (Fig. 11). The fissure eruptions formed a thin platform with an almost constant elevation of 230 m. The platform does, however, slope to the south at 0.75 m/km and is at an altitude of 210 m at a distance 30 km to the south. The maximum thickness of the basalt is 40 m, although most frequently it is less than 20 m thick and there are individual stacks of lava that do not exceed 2 m in thickness. Lacombe separated the volcanic constructs into five distinct groups: (1) Simple scoria cones—these have basal diameters of less than 3 km and are as much as 150 m in height. (2) Horseshoe craters—these have been opened by erosion. (3) Crater cones—these are sparse except where subterranean water infiltrated without destroying the rampart. (4) Explosion craters. (5) Rift valleys and volcanotectonic depressions.

Specifically, Lacombe mentioned Ya Lom, which has a crater 60 m deep and 800 m in diameter; Lom Kot, which is a crater 600 by 300 m associated with a fracture in the sandstones and alluvium of the plain of Thong Samoth; Ya Om–Ya Ra, which are a pair of crater lakes 200 m in diameter; and Bo Kham, which is a large, well-preserved edifice. There are composite cones in the Ya Lom complex, with 12 cones and craters in a region of 10 km. The flows include olivine basalts and both saturated and undersaturated alkaline basalts. The edifices are composed of alkaline basalts, mottled sodic alkaline basalts, and differentiated sodipotassic rocks (trachyandesites with oligoclase). Volcanic materials include welded scoria, breccia, vesicular basalt, glassy basalt dikes, lapilli tuffs, tears, tektites, and bombs of peridotite.

The plateau was formed at the end of one of the old Quaternary epochs and has been correlated with the 100-m platform of central Indochina. Activity continued during the Middle and recent Quaternary. Activity commenced with fissure eruptions before 0.7 m.y. ago and appears to have ceased rapidly about 0.7 m.y. ago. After erosion attacked the early flows, new eruptions took place on the plateau forming the upper flows. These overflowed to the north and west to lower altitudes (180 to 200 m). The following period was marked by more explosive eruptions, which created the numerous small edifices. The final stage consisted of the opening of many gaseous explosion craters and the emission of great quantities of cinders and pumice. The location is around 13.5°N, 107.2°E.

87. Haut Song Be (Haut Chhlong). This is a basalt plateau with dimensions of 100 km northeast-southwest and about 50 km northwest-southeast, extending into Vietnam (see listings

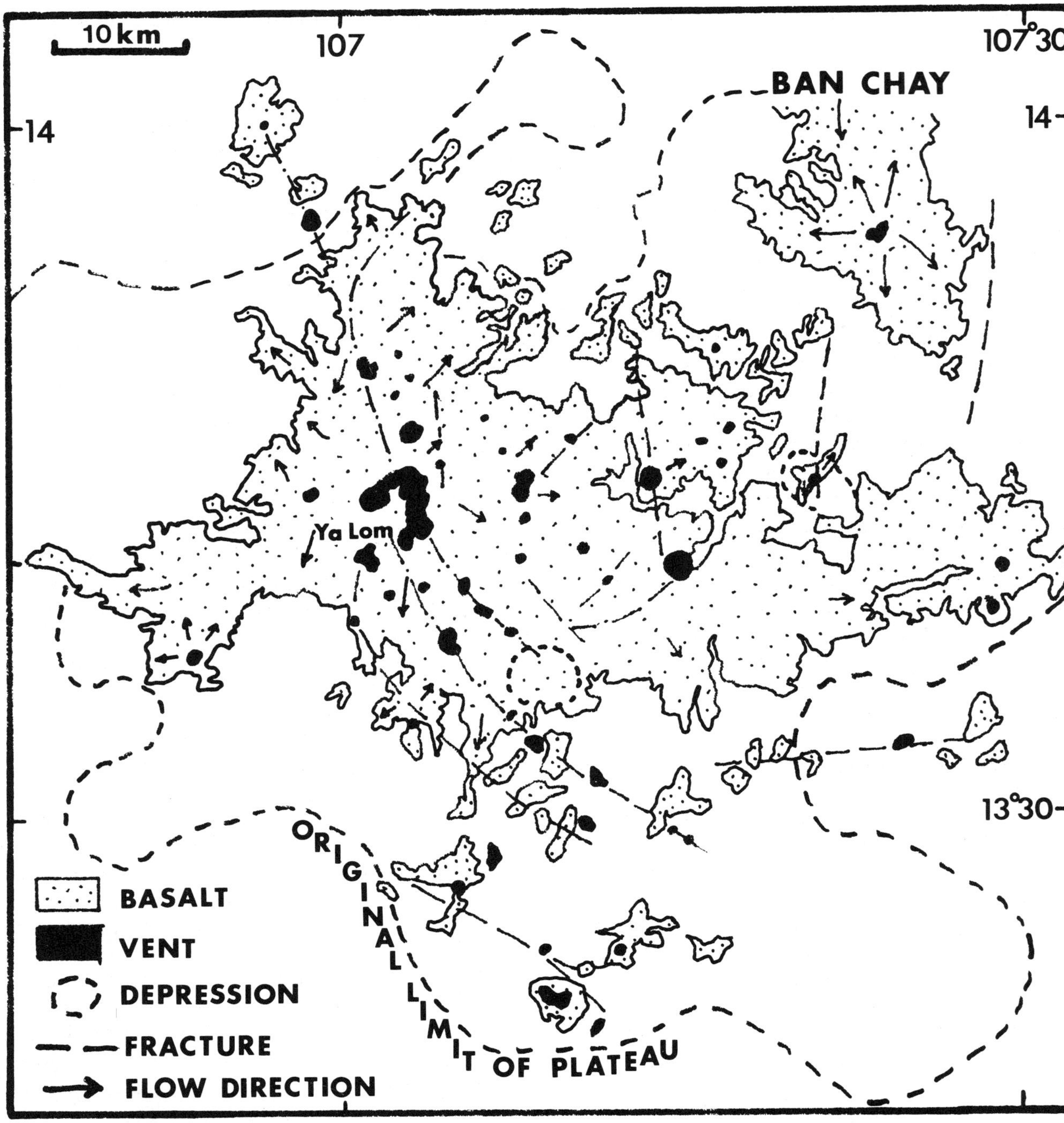

Figure 11. Distribution of volcanic materials on Bokeo (Ratanakiri) Plateau (listing 86) of Kampuchea (Cambodia). (Modified after Lacombe [1969].)

102 and 103). The flows are associated with transverse fractures (Saurin, 1967). The location is centered around 12°N, 107°E.

88. Snoul. Lower Pleistocene–age basalts with dimensions of 10 by 25 km occur to the west of the town of Snoul (Saurin, 1967). An area of basalt 50 by 25 km also surrounds the town of Mimot. These are tholeiitic lavas (Barr and Macdonald, 1981). The location is at 12.1°N, 106.4°E and 11.8°N, 102.2°E.

89. Chup. Lower Pleistocene–age basalts occupy an area 50 by 10 km. These appear to be associated with the same west-northwest–trending fracture that is found at Mimot (Saurin, 1967). The location is at 11.9°N, 105.7°E.

90. Cham (Prek Kak). This is an expanse of basalt 50 by 50 km separated from the Chup plateau by the Mekong River. The basalts are also associated with the same west-northwest–trending system. The eruption took place along the central axis of an upwarp that affected the entire area (Carbonnel and others, 1972). Lavas form the 80- to 100-m terrace of the Mekong river. The basalts include nepheline hawaiite, tholeiite, and basanite (Barr and Macdonald, 1981). The location is at 12.1°N, 105.6°E.

91. Pailin (Cardamomes). Limburgites cover an area of 4.0 by 0.8 km and were erupted along the flanks of an uplift associated with a northwest-trending fracture. Lavas were deposited on the 10-m river terrace. These form part of the earliest series of Pleistocene lavas in the area, being dated at 1.99 ± 0.19 m.y. (Carbonnel and others, 1973). The hills Phnum O Tang, Phnum Ko Ngoap, and Phnum Yat are considered to be the eroded remnants of volcanic cones. K-Ar ages include 1.09 ± 0.13 m.y. for Phnum O Tang, 1.43 ± 0.10 and 1.42 ± 0.08 m.y. for Phnum Ko Ngoap, and 1.3 ± 0.17 m.y. for the O Chra volcanic pipe. These rocks are nepheline hawaiites and basanites. An older fission track age of 2.42 ± 0.42 m.y. was obtained on a zircon from a hawaiite at Phnum Yat. Thus there appear to have been several phases of volcanism (Barr and Macdonald, 1981). Like the nearby area of Chanthaburi in Thailand (listing 84), the Pailin basalts are associated with gem-quality zircons and rubies. It is also the most important sapphire deposit in the world. Nearby basalts at Bar Yakkar cover 3.5 by 0.8 km and were deposited on the 25-m river terrace (Carbonnel and others, 1973). The location is at 12.9°N, 102.6°E.

92. Plateau des Bolovens. This is a large expanse of basalt approximately 75 km in diameter at an elevation of 1,000 to 1,200 m, with flows occupying the valleys that radiate from the plateau center. The basalts are associated with north-northeast–trending fractures (Saurin, 1967). A zircon has yielded a fission track age of 1.10 ± 0.9 m.y. (Carbonnel and others, 1972). The location is at 15°N, 106.3°E.

VIETNAM

There are extensive basaltic flows within Vietnam, but the information concerning them is often poor or located in journals that are not readily obtainable. For example, "Ils se rétrouvet aussi dans les Valles, commes aux environs de Cong Son et, près de Song Cau, forment des coulée qui descendent jusqu'au niveau de la mer." (They are found again also in the valleys, as around Cong Son and, near Song Cau, forming flows which descend almost to sea-level; Saurin, 1936). The basalts in southern Vietnam appear to be related to the same tectonic activity associated with the basalt plateaus of Laos and Cambodia.

93. Unnamed. This is a group of four basaltic cinder cones of Pleistocene age located at approximately 19.2°N, 105.7°E (IAVCEI, 1973).

94. Unnamed. This is a group of three basaltic cones of Pleistocene age located immediately south of the cones described in listing 93 (IAVCEI, 1973).

95. Vinkh-linkh. This is a Pleistocene basaltic cone within a basalt field covering approximately 40 by 10 km (IAVCEI, 1973). The cone lava is a hawaiite. The surrounding lavas have an extensive lateritic cover and probably represent an earlier episode of volcanism (Barr and Macdonald, 1981). The location is at approximately 17°N, 107°E.

96. Cu-Lao Bo Bai. A destroyed basaltic cone and caldera (a maar?) rising to a height of 20 m (26 m above sea level). This probably erupted within the last 2,000 years (IAVCEI, 1973). The location is at 15.43° N, 109.06°E.

97. Cu-Lao Re Island. There are three basaltic cones with heights of 20, 74, and 130 m. Two of them form a well-preserved double cone and the third is a "caldera." It is also possible that these have erupted within the last 2,000 years (IAVCEI, 1973). The location is at 15.38°N, 109.1°E.

98. Toroeng Prong. This is a basaltic volcano 60 m high with a "caldera" containing a crater lake and three "mouths" (IAVCEI, 1973). The location is at 14.9°N, 108°E.

99. Pleiku-Bantour area. This is an extensive area of basaltic lava some 75 by 75 km in size with a number of volcanic constructs, most of which are in a poor state of preservation. The IAVCEI (1973) catalog lists five calderas: Plei-monu (80 m high), Chi-haedrong (Chi Hodron) mount (122 m), La Bang mount (55 m), Pleiku (60 m), and Chi-ten (70 m). There are also two basaltic lava cones; Chi-holam mount (60 m) and Pleitott mount (40 m). The younger cones are more alkalic than the older lavas and include alkali olivine basalts and hawaiites, whereas the older flows are hawaiites and tholeiites. The lavas have surface reddish-brown latosols several meters thick. A hawaiite from Pleiku had a K-Ar age of 2.1 ± 0.1 m.y. (Barr and Macdonald, 1981F). The location is at 14°N, 108°E.

100. Ban Chay. This is a basalt area of approximately 200 km^2 situated to the immediate northeast of Bokeo (86) at approximately 14°N, 107.5°E. Lavas appear to have flowed radially away from a vent area located in the south-central portion of the field (see Fig. 11) (after Lacombe, 1969).

101. Darlac. This is the southern extension of the Pleiku field. A sample collected from northeast of Ban Me Thuot was a hawaiite that yielded a K-Ar age of 3.4 ± 0.2 m.y. (Barr and Macdonald, 1981). The location is centered on 12.8°N, 108.4°E.

102. Bu Jerman. This is a basaltic shield volcano, 40 m in height, situated within the same field as the "caldera" in listing 103. The volcano is prehistoric in age (IAVCEI, 1973). The location is at 12.15°N, 106.8°E.

103. Bu Kong Rong. A badly preserved basaltic "caldera" some 20 m in height and of prehistoric age (IAVCEI, 1973), this volcano is located within an extensive lava field some 150 by 70 km in size called Haut Dong Nai and Djiring Plateau. It and the Haut Chhlong Plateau to the west (see listing 87) have an elevation of 800 to 1,000 m. The basalt is deeply weathered to a lateritic red earth; the only preserved landforms are located in the east, where they are found as isolated cones and flows. The eastern flows are considered to be younger than those in the west. Samples collected from the Djiring Plateau were mainly tholeiites but also included a mugearite and a hawaiite. These samples yielded relatively old K-Ar ages of 8.8 ± 0.3, 13.3 ± 1.1, and 11.6 ± 0.3 m.y. (Barr and Macdonald, 1981). Isolated outcrops to the east of the main Djiring Plateau are basanitoid. The location is centered around 11.6°N, 108.2°E.

104. Lien Khang Mount. This is a well-preserved basaltic "caldera" 10 mi high. Of pre-historic age (IAVCEI, 1973), it also lies within the Djiring Plateau. See Plate 307 in Short and others (1976).

105. Suzanna Mount. This is a well-preserved basaltic lava cone rising 180 m above its base (392 m above sea level) and is of prehistoric age (IAVCEI, 1973). The location is at 10.83°N, 107.35°E.

106. Cam Tiem Mount. This lava cone is well-preserved, although it is of prehistoric age; it rises to a height of 100 m (IAVCEI, 1973). It is located at 10.81°N, 107.36°E.

107. Nui Dat. A basaltic pyroclastic cone 77 m high in a poor state of preservation and of pre-historic age, this structure is located at 10.53°N, 107.18°E. This and the structures described in listings 105 and 106 are located on the Bas Dong Nai lava field, which surrounds the town of Xuan Loc. Basalts from here have been dated at 0.65 m.y. old by the fission track method (Carbonnel and others, 1972). K-Ar dating on a hawaiite produced an age of 2.6 ± 0.2 m.y. (Barr and Macdonald, 1981). Basanite and hawaiite are also found on the Xuan Loc Plateau. Saurin (1968) described a section 5 km west of the town as being composed of 2 m of vesicular basalt that overlies 3 m of compact basalt, 3 m of red earth mixed with basalt, 14 m of compact basalt, 4 m of scoriaceous basalt, and finally, 74 m of lapilli tuff and pumice. Gems, mainly zircon, are mined at Xuan Loc. The gems are found in gravel deposits derived by rapid weathering of the younger vesicular basalts. The composition of the parent basalt is not known (Barr and Macdonald, 1981). Xuan Loc is located at 11°N, 107.2°E.

108. Bas Song Be. This is a comparatively small area of basalt approximately 25 km in diameter that lies immediately to the southwest of the much larger Haut Song Be (Haut Chhlong) (see listing 87). Samples collected from Bas Song Be are tholeiites (Barr and Macdonald, 1981).

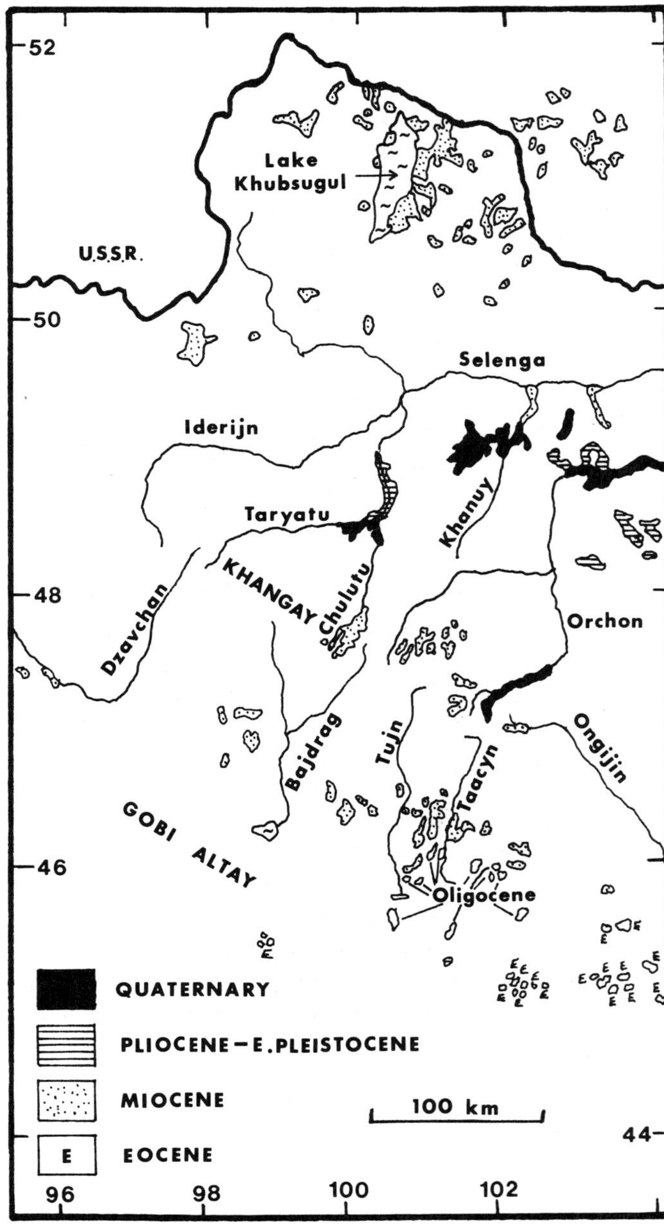

Figure 12. Distribution of volcanic materials in north-central region of Mongolia. (Modified after Devyatkin and Smelov [1979].)

MONGOLIA

Many of the Mongolian volcanic deposits have been examined only recently by Soviet and Mongolian geologists, as many lie in isolated and desolate regions. The majority of the Cenozoic deposits described thus far lie in isolated areas in the center of the country; there may, however, be undescribed deposits in the east of the country that are extensions of the Chinese volcanic belts. A location map for the central Mongolian volcanics is presented in Figure 12.

109. Nomogan-Ula and Khutza-Ula. Two basaltic volcanoes Pliocene to Lower Pleistocene in age, rise 21 m and 162 m, respectively, above their bases (IAVCEI, 1973). In this same area, Tikhonov (1977) noted the existence of 18 ring structures aligned parallel with the Dzuunbayan basin system. At the periphery of the rings there is a ridge ranging from several hundred meters to several kilometers in width. Inside the ridges are depressed zones, which appear to be subsidence calderas. In the middle of nearly all the structures is a conical hill as much as 3 to 5 km in diameter. The more ellipsoidal structures tend to have two conical hills. The ring structures are chiefly Mesozoic extrusives, but the central cones are, in some places, small Cenozoic volcanoes. The Mount Han Bogdo structure is a Permian granite intrusion measuring 30 by 45 km, within which there is ridge 20 to 22 km in diameter of intermediate extrusives 300 to 400 m wide. The area inside the ring is depressed like a subsidence caldera. The location is approximately 42.5°N, 107.5°E.

110. Tabun Urtu-Ula. This is a Pleistocene-age volcano rising about 67 m above its base and 1,167 m above sea level. The location is at 44.35°N, 107.63°E (IAVCEI, 1973).

111. Middle Gobi. These are three "destructive" cones situated along an east-west fault, located 45° to 45.58°N, 105.83° to 107.58°E. The cones are basaltic and possibly of Upper Pleistocene to Holocene age. They are 60 to 80 m in height (IAVCEI, 1973).

112. Unnamed. These are three volcanoes of Pliocene to Lower Pleistocene age, ranging in height from 31 to 167 m above their bases (IAVCEI, 1973). The location is 42.26° to 42.38°N, 105.18° to 106.76°E.

113. Unnamed. These two "destructive cones" of possible mid-Pleistocene age are composed of basalt. They rise above 50 m in height (IAVCEI, 1973). The location is 48.16°N, 109.83°E.

114. Orchon Gol (Orckon Gol). Several flows of Pliocene basalts form the 90-m terrace along the Orchon River, 10 km below Orchon Somon. The age range varies from 6.4 ± 0.5 m.y. in the lower part to 3.1 ± 0.2 m.y. in the upper. Upriver, opposite the Dulan-Khara volcano, the same complex forms the lower part of the 120-m terrace. The two to three flows at this locality are thick columnar basalts with sign-variable magnetization and have been dated at 4.0 ± 0.2 m.y. in the upper part. Lower Pleistocene basalts form the upper part of this terrace. These latter basalts comprise 10 to 12 finely stratified flows, all reversely magnetized, and are dated at 1.5 ± 0.1 m.y. North of the Orchon valley, the Lower Pleistocene basalts that fill in the lower portions of the watersheds are reversely magnetized and have been dated at 1.8 and 2.1 ± 0.1 m.y., perhaps belonging to the Matuyama epoch. These basalts are associated with the Bayan-Tsagan-Dzurikhe crater. At Orchon Somon, the Lower Pleistocene basalts form the 20- to 25-m terrace, are normally magnetized, and have been dated at 1.8 ± 0.1 m.y. (Gilsa-Olduvai?) (Devyatkin and Smelov, 1979). The volcano Bus Obo (47.11°N, 109.08°E), described as a 90-m-high basaltic "destructive cone" of possible Upper Pleistocene to Holocene age (IAVCEI, 1973), occurs in this general area; its relation to the two previously described volcanoes is not readily apparent.

115. Ikha-togo-ula and Bogo-togo-ula. These are two Quaternary-age basaltic volcanoes rising 219 and 28 m, respectively, above their bases (IAVCEI, 1973). The location is at 47.88°N, 103.08°E.

116. Khanuy Gol (Chanuj Gol). Miocene basalts occur as a residual of the 100-m terrace along the lower reaches of the Khanuy river. The basalts have been dated at 12.5 ± 1.0 m.y. and are normally magnetized. Pleistocene basalts form the 10- to 12-m terrace in the middle part of the Khanuy valley associated with two volcanoes (115?). There is also a small field to the north of the Bulgan settlement. Devyatkin and Smelov (1979) described the latter area as containing semi-destroyed volcanoes and the beautifully preserved Togo cone (115?). This could be the same area described in the IAVCEI (1973) catalog as being a group of 10 lava cones of Upper Pleistocene to Holocene age, ranging from 28 to 186 m in height and composed of olivine basalt. The location is at 48.31° to 49°N and 102.08° to 103.4°E.

117. Khubsugul region (Chovsgol). Basalts dated as Pliocene to Quaternary in age form large lava flows and are the remnants of a single volcanic plateau surrounding Lake Khubsugul. The flows are multi-layered with 6 to 10 individual units, totaling 100 to 150 m in thickness. The basalts of the eastern Khubsugul can be traced northeastward to the Khamar Daban range of the Soviet Union (Devyatkin and Smelov, 1979). The IAVCEI (1973) catalog lists two unnamed volcanoes at 51.33°N and 99.33°E as being of basaltic composition and Upper Pleistocene age. These volcanoes are situated along a north-south fault.

118. Unnamed. These are five cone-shaped, basaltic volcanoes of Pleistocene age ranging in height from 30 to 100 m. They are located from 44.03° to 44.45°N, 103° to 104.06°E (IAVCEI, 1973).

119. Unnamed. These are five basaltic volcanoes of Pliocene to Lower Pleistocene age, ranging from 13 to 192 m in height. They are located from 43.28° to 43.5°N and 103.28° to 104.73°E (IAVCEI, 1973).

120. Durbulzhi-ula. A 45-m-high basaltic volcano of Pliocene to Lower Pleistocene age, this structure is located at 42.8°N, 102.08°E (IAVCEI, 1973).

121. Ubur-khangay. Six basaltic cone-shaped volcanoes of Pleistocene age range in height from 96 to 235 m (IAVCEI, 1973). They are located between 44.25° to 44.5°N and 101.33° to 102.3°E.

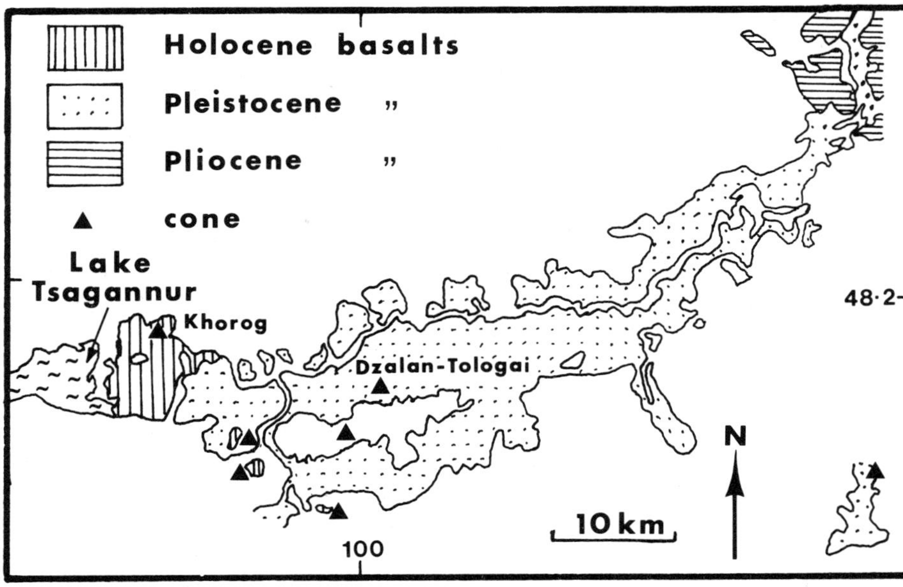

Figure 13. Distribution of volcanic materials in Taryatu Depression (listing 122) of north-central Mongolia. (Modified after Kepezhinskas and others [1975].)

122. Taryatu-Chulutu area. Miocene basalts are recorded as forming the highest terrace (130 to 200 m) on the Chulutu River; Pliocene basalts form the 100- to 120-m terraces (Figs. 12, 13). Below Mauren, the upper third of the Pliocene section comprises a finely stratified sequence of thin flows 3 to 6 m thick, whereas the lower part is formed of basalt sheets 15 to 20 m thick, which are cut by sills, dikes, and stocks. The stocks reach 100 m or more in diameter (Devyatkin and Smelov, 1979). The upper flows are pyroxene trachybasalts and trachytic andesite basalts, whereas the lower flows are potassic limburgites and augitites. The lower flows also contain xenoliths of sanidine, peridotite, and spinel-lherzolite (Kepezhinskas and others, 1975). Lower Pleistocene basalts form the 60-m terrace of the Chulutu River.

Pleistocene basalts form the 40- to 60-m terrace of the Taryatu basin. The flows have normal magnetization and have been ascribed to the Brunhes epoch (Devyatkin and Smelov, 1979). The Pleistocene sequence ranges from 40 to 90 m thick and comprises potassic basanites in the lower part and limburgites in the upper; again both are associated with ultrabasic xenoliths (Kepezhinskas and others, 1975). The Holocene basalts occur mainly within the western part of the depression and are associated with six cinder cones (Devyatkin and Smelov, 1979). The IAVCEI (1973) catalog described about 10 volcanoes ranging from 60 to 170 m high; of olivine and analcime basalt, they occur along the side of a graben-like valley located from 48° to 48.33°N and 99.1° to 100.25°E. The Holocene flows are 3 to 5 m thick and comprise leucite basanites, leucite tephrites, potassic hawaiites, and augitites (Kepezhinskas and others, 1975). The youngest flows contain the largest quantity of ultrabasic xenoliths and have the highest alkali content (K_2O reaches 3 to 5 weight percent) (Devyatkin and Smelov, 1979). Two named volcanoes associated with these flows are Khorog and Dzan Tologai (see Appendix 6 for dimensions). Khorog is breached to the south at the location of a former lava tube. The cone is composed of cemented agglutinates, scoria, horizons of welded bombs, and huge blocks (3 to 5 m in diameter). An ash train, 3 to 5 m thick and containing scoriaceous lapilli and bombs up to 1 m in size, stretches to the east of the volcano (Devyatkin and Smelov, 1979).

123. South Khangay Plateau. This includes the areas between the Shand, Tatsin, Baydrag, and Dzabkhan Gols (alternately spelled Hsanda, Taacyn and Tujn, Bajdrag, and Dzavchan) (Fig. 12). Volcanics include limburgites, analcime andesite basalts, trachybasalts, and sanidine and biotite basalts distinguished by a high K_2O content (average, 3.14 weight percent) (Devyatkin and Smelov, 1979). Oligocene basalts include a 10-m-thick layer along the Dzabkhan River dated at 31 ± 2 m.y., a "lower basalt layer" on the Shand Gol-Tatsin Gol interfluve with ages ranging from 28 to 32 m.y.; an "upper basalt layer" along the Tatsin Gol dated at 24 ± 1 m.y.; and a series of five closely spaced basalt layers with an overall thickness of 30 m near Mt. Dushu-sharoula, the upper layer of which has been dated at 21 ± 2 m.y. More fragments of "watershed" plateau basalts, up to hundreds of meters thick, cover large areas of southern Khangay. A date of 28.5 ± 1 m.y. has been obtained for a flow near the middle of this sequence (Devyatkin and Smelov, 1979). Some basalts, possibly representing this sequence, were mapped by Berkey and Morris (1927).

Miocene basalts are offset slightly to the north of the Oligocene basalts. Basalt horizons along the Tujn and Tatsin Gols have been dated as being between 16 and 20 m.y. in age. The basalts form residual plateaus, are multi-layered, and exhibit columnar jointing. Several volcanic necks are associated with this sequence and vary from a few hundred meters to 1.5 km in diameter and are up to 120 m in height. Devyatkin and Smelov (1979) mentioned two necks; Mt. Dushu-sharo-ula and Kholbo-teg-ula. Basalts from the latter have been dated at 18.5 ± 1 m.y. in age.

124. Gobi Altai. Eocene-age necks extend in a chain from Bayan-leg Somon eastward. Mount Khaton-sudal (a typical neck) has a basal diameter of 200 m, a height of 100 m, and a summit diameter of 50 m. It is composed of columnar basalt dated at 57.0 ± 2 m.y. In the foothills of the Arts-bogdo Range, 30 to 35 necks are arranged in bunches of three to four in a chain along a latitudinal fault. These range up to 200 m in height and have basal diameters of tens to hundreds of meters. Northwest of Bulgan Somon, such as at Mount Barun-tabun-khayrkhan-ula, the necks contain flows. One flow at the latter neck has been dated at 47.0 ± 2 m.y. Volcanic rocks within the Altai include limburgites, augitites, analcime basalts, trachybasalts, and trachyandesites with an average K_2O content of 1.72 weight percent (Devyatkin and Smelov, 1979) (see Appendix 7). Leucite basalts were described by Kepezinskas and others (1975) as occurring along the Bulgan fault zone between the Altai Range and the Barun-khurai depression. Florensov and Solonenko (1963) have described Dziran as a Tertiary shield volcano located at approximately 44.75°N, 100.75°E. It is composed of trachybasalts and trachydolerites that are inclined 5 to 7° away from the center of the structure and cover an area of 200 km^2.

125. Dariganga. This widespread area of basaltic volcanic rocks is an extension of the inner Mongolian field. The IAVCEI (1973) catalog stated that the area contains more than 200 lava and cinder cones of Quaternary age, ranging in height from 25 to 300 m, that are situated along northeast-trending parallel fissures. Devyatkin and Smelov (1979) stated that some of the volcanism was Pleistocene; they also noted that some of the cones were normally magnetized (e.g., Lun-ula, Dusy, Shilin-bodgo), whereas others were reversely magnetized (e.g., Akhat, Barun-nerte-ula, Dzunnerette, Bayan-tsagan). The lavas comprise picritic basalts, limburgites, analcime-, andesine-, and oligoclase-basalts, and trachybasalts, with an average K_2O content of 1.72 weight percents. Dariganga was described by Genshaft and others (1978) as a sodic province as distinct from the potassic province of north and central Mongolia. They also noted that the southern part of Dariganga is richest in xenoliths (lherzolites, wehrlites, dunites, harzburgites, and pyroxenites) and that basalts dominate in the north and cinders in the south (see also listing 28).

SOVIET UNION

Volcanic activity in the eastern Soviet Union, excluding the Kamchatka and Koryak Peninsulas, has gradually declined since the early part of the Cenozoic. Takahashi (1983) has recently reviewed the late Mesozoic to early Cenozoic magmatic evolution of this region. The Cenozoic volcanic activity has been largely confined to the eastern sea coast and along the southern border (Fig. 1b).

126. Upper Anadyr River. In the upper Anadyr volcanic trough, the Upper Cretaceous volcanics are covered by a gently sloping Paleogene basalt that is not more than 900 m thick. The Kuiviveem complex comprises three intersecting calderas, 55 km in diameter, within which there are three main extrusion centers. According to geophysical data, the center of the complex is filled with at least 2,000 m of extrusives (ignimbrites). Some 800 km^3 of material has been erupted from the volcano. The Pyrkhanaivaam and Varynnaiskaya calderas, both 20 km in diameter and intersecting, erupted 120 and 100 km^3 of ignimbrite material, respectively. The Gremuchaya caldera (approximately 25 km in diameter) has been bisected by the Okonaito-Yuomkuveen fault. Also within the trough are four named Paleogene shield volcanoes. The Shchitovaya volcano is basaltic, 25 by 70 km in size, and erupted 430 km^3 of material. The Marakveem, Terpukhoi, and Kyrgannivaam shields have diameters of 50, 55, and 60 km, respectively (after Zhupakhin and others, 1977). Twenty-five eruption centers are shown on the map by Zhupakhin and others (1977); see Figure 14. The location is centered around 65°N, 168°E.

127. Aluchin Group. This consists of extensive sheets of lava and three cones (Fig. 15). The cones have been called Alpha, Beta, and Gamma (see Appendix 6). Alpha is a double cone that alternated in eruption between quiet effusion and explosive events; Beta is similar but is partially destroyed; and Gamma is a pile of semi-consolidated agglomerate. The three volcanoes are located along a fault zone. The lavas from these vents infill a valley, varying in width from 2 to 8 km and extending some 35 km north-south. They occupy a total area of 250 km^2 and have a volume of 0.05 km^3. The lavas vary in thickness (2 to 40 m) both downvalley and across valley. The Beta and Gamma flows were apparently produced in three cycles and predated the Alpha flows. The flows are basalts and trachybasalts close to tephrite in composition although lacking feldspathoids. The volcanoes are estimated to have formed around 1000 AD (Gorodinskiy and others, 1967). The location is approximately at 66.1°N, 165.4°E.

128. Anyuy (Anjuisky) Group. Early eruptions in this area were of trachybasalts from a series of fissures totaling 15 to 17 km in length, along the valley floor. There were at least three flows, which have combined dimensions of 2 to 4 by 50 km, an average thickness of 30 m, a coverage of 100 km^2, and a volume of 3 km^3. The lava is asymmetric in thickness across the valley, being thickest (50 m) on the east side close to the fissures. Small lava lakes and tubes formed in these flows. Aniusk is a small cone (see Appendix 6) that experienced explosive eruptions and small outpourings of trachybasalt flows. One flow has a length of

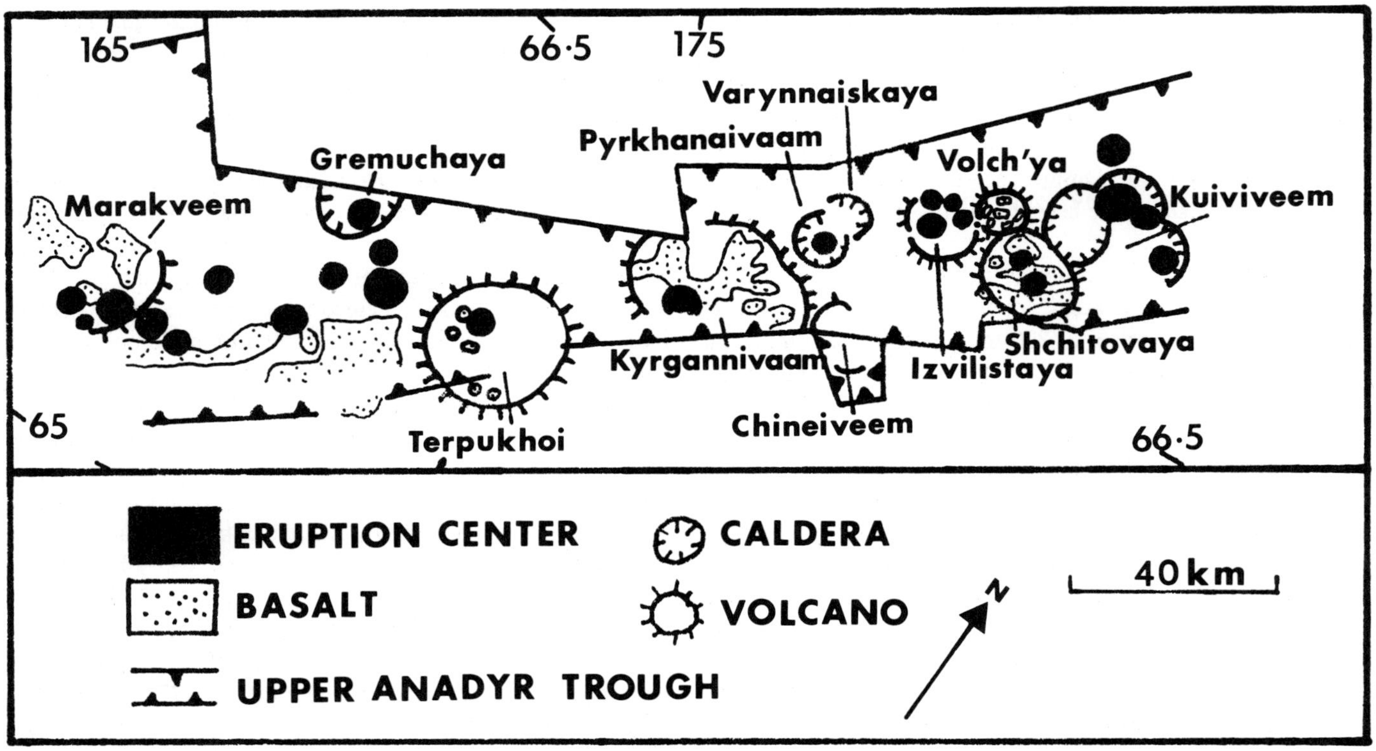

Figure 14. Distribution of volcanic materials in Anadyr Trough (listing 126). (Modified after Zhupakhin and others [1977].)

16 km, an average thickness of 20 m, a coverage of 24 km^2, and a volume of 0.48 km^3. The eruptions increased in basicity with time; the volcano was probably active in the 14th and 18th centuries (Ustiev, 1959a). Another crater with a small flow was later discovered to the southwest of Anyuy, a distance of some 2 km (Chasovitin, 1963). The location is at approximately 67.7°N, 165.2°E.

129. Bilibina. This is a cinder cone (see Appendix 6 for dimensions) that alternated between eruptions of small quantities of very fluid lavas and explosions. The lavas are andesite and tephrite basalt of aa form. The volcano, situated in a structural valley, is older than the Aluchin Group, maybe Lower Pleistocene in age (Chasovitin, 1963). The location is at 66.33°N, 167.5°E.

130. Unnamed. This is a volcano of Lower Pleistocene age composed of andesite basalt, located at 68.16°N, 167.45°E (IAVCEI, 1973).

131. Unnamed. This is a volcano of Lower Pleistocene age composed of trachybasalt, located at 67.48°N, 170.28°E (IAVCEI, 1973).

132. Balagan-Tas (Indighirsky?). This is a cone-shaped volcano with a poorly preserved crater (see Appendix 6). It occupies an area of 1.8 km^2, with lavas covering 4 to 5 km^2. The lavas probably had a maximum thickness of 10 m, are subalkaline basalts in composition, and are dated as early Quaternary (Argunov and Gavrikov, 1960). The IAVCEI (1973) catalog, however, suggested that Balagan-Tas (located at 66.43°N, 143.73°E) might be the volcano Indighirsky, which is supposed to have erupted in the 1770s. Ustiev (1959b), however, seemed to place Indighirsky farther south of the location of Balagan-Tas, near the town of Zashiversk.

133. Majak. This is a Quaternary-age dome composed of liparite, situated a little to the north of Balagan-Tas at 66.45°N, 147.73°E (IAVCEI, 1973).

134. Unnamed. This is a group of basalt and andesite-basalt volcanoes situated along a fault along the River Indighirka at approximately 69.16°N, 147.15°E. The volcanoes are of Lower Pleistocene age (IAVCEI, 1973). Ustiev (1974) mentioned the volcano D'yakhtardakh in the Polousnyy Mountains as being composed of trachytes and liparites, and Mt. Polevaya in the Ulakan Cis range, as being composed of early dacite, liparite, and trachyandesite with minor andesite basalts and basalt.

135. Other volcanics in the northeast Soviet Union. Ustiev (1959b) described further volcanics including basaltoids in the basin of the River Pravaia Yasachnaia, a tributary of the

Kolyma; Quaternary basaltoid lavas overlying glacial deposits in the northern Taigonos Peninsula; ash blankets in the Magadan region; Quaternary lavas in the Malyi Anuiy valley; and a crater lake of Quaternary age—Elgytkhyn—in the Anadyr uplands. The latter crater lake is now considered to be a meteorite impact crater (Alyunin and Dabizha, 1980; Gurov and others, 1979) with an age of 3.5 ± 0.5 m.y. Chasovitin (1963) also showed seven phreatic explosion funnels to the southeast of Chaun Bay and three on the northwest side of the bay. He also showed nine additional funnels along the lower reaches of the Kolyma and Bol Anyuy Rivers but did not elaborate on their origin.

136. North coast of the Sea of Okhotsk. Plateau basalts came up along large faults at the end of the Cenozoic period, sometimes accompanied by small volumes of ignimbrites (Speranskaya, 1966). These basalts overlie some 800 m of Cretaceous acid volcanics.

137. Mt. Belaya. This is a trachyliparite dome of possible Pliocene age. This and several other domes located farther to the south range in height from 50 to 150 m (Plakhotnik, 1968). The location is at approximately 54°N, 140.1°E.

138. Mt. Lysaya. This is a trachyliparite dome of possible Pliocene age located at approximately 52.5°N, 140.3°E (Plakhotnik, 1968).

139. Mt. Castri. This is another dome of Pliocene age. There are 11 other domes of similar age in this area, comprising Mounts Kopna, Sopka, Konus, Shapka, Nevel'skogo, Arbat, and five other unnamed mounts. Their compositions range from trachyliparite through andesite and dacite (Plakhotnik, 1968). The location is at approximately 52°N, 142.2°E.

140. Shishlovsky. This is a 40-m-high cone of Pliocene age composed of basalt and basaltic tuff located at 51.38°N, 140.13°E (IAVCEI, 1973).

141. Uysky. This is a shield volcano of Pliocene age, which reaches a height of 300 m above its base. This edifice is composed of subalkaline basalt and tuff (IAVCEI, 1973), and the caldera contains a dacite dome (Plakhotnik, 1968). The location is at 49.03°N, 140°E.

142. Bolonsky. This is a 15-m-high cone-shaped volcano composed of leucite basalt and tuff of Pliocene age. The location is at 49.8°N, 136.26°E (IAVCEI, 1973).

143. Sindinsky. Similar to Bolonsky except that it is 32 m high and composed of alkaline basalt and tuff, this structure is also of Pliocene age. The location is at 48.08°N, 136.16°E (IAVCEI, 1973).

144. Sviatogorsky. A cone-shaped volcano of alkaline basalt, this structure reaches 68 m in height and again is of Pliocene age. The location is at 47.86°N, 135.41°E (IAVCEI, 1973).

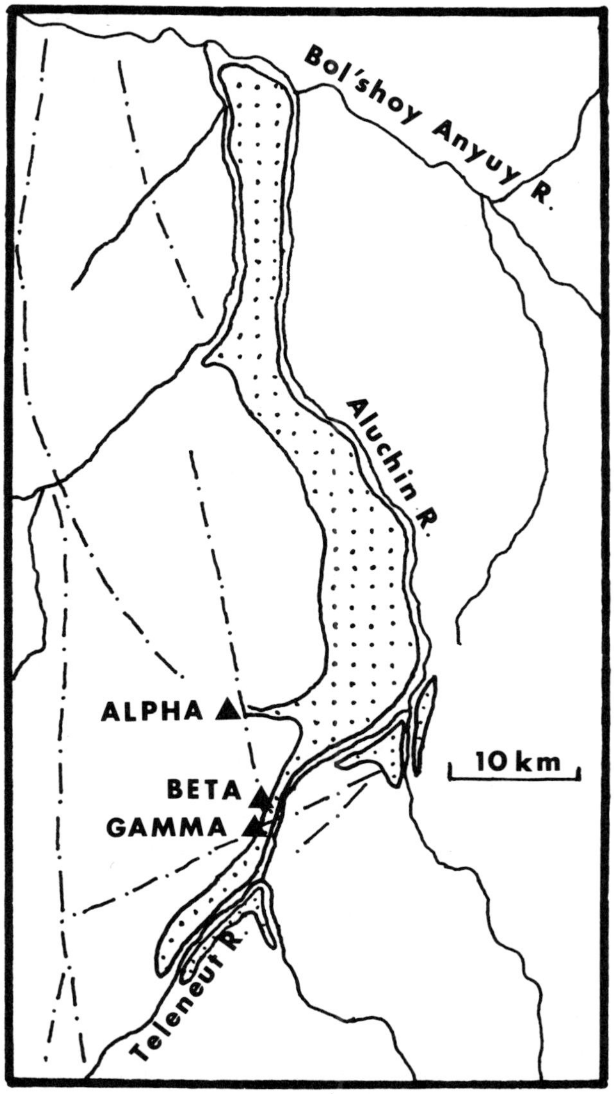

Figure 15. Cones (triangles) and lava flows (dotted) of Aluchin region (listing 127) of northeast Siberia. Dot-dash lines represent fault traces. (Modified after Gorodinskiy and others [1967].)

145. Imansky. A 56-m-high cone-shaped volcano of Pliocene age, this structure is composed of nepheline basalt and tuff. The location is at 45.96°N, 133.76°E (IAVCEI, 1973).

146. Medvegiy. Also composed of nepheline basalt but having a domical form, this structure reaches a height of 247 m above its base and is of Pliocene age. The location is at 45.2°N, 133.46°E (IAVCEI, 1973).

147. Ostry. This is a 100-m-high dome-shaped volcano of Pliocene age composed of nepheline basalt. The location is at 45.13°N, 133.46°E (IAVCEI, 1973).

148. Sviaghinsky. This shield volcano is composed of alkaline basalt and tuffs of Pliocene age and reaches a height of 30 m above its base. The location is at 44.83°N, 133.2°E (IAVCEI, 1973).

149. Merkushovsky. A somewhat larger shield volcano with a height of 156 m, this structure is again of Pliocene age and is composed of alkaline basalt and tuff. The location is at 44.36°N, 132.8°E (IAVCEI, 1973).

150. Baevsky. A 113-m-high cone-shaped volcano of Pliocene age, this structure is composed of alkaline basalt and dolerite. The location is at 43.91°N, 131.66°E (IAVCEI, 1973).

151. Razdolninsky. This is also a cone-shaped volcano of Pliocene age composed of basalt; it is 258 m high. The location is at 43.5°N, 131.91°E (IAVCEI, 1973).

152. Southern Sikhote-Alin. There was extensive volcanic activity in Sikhote-Alin throughout the Cretaceous and Tertiary periods. Four Tertiary cycles have been recognized (Bersenev and others, 1966); Paleocene, Eocene-Oligocene, Miocene, and Pliocene to Recent. The earlier Senonian-Danian cycle was marked by acid extrusives, whereas the Cenozoic cycles were acid/intermediate rocks: 50/50 in the Paleocene, thin basic and acid rocks in the next cycle, followed by mainly basic extrusives with very little acidic material, and finally entirely basaltic material in the Pliocene to Recent. The "tufflavas" or ignimbrites of the second cycle cover considerable areas (up to 40 km^2) and in places reach thicknesses of 300 to 500 m.

153. Northeast Udokan Plateau. The IAVCEI (1973) catalog listed a group of nine Holocene-age basaltic cinder cones ranging in height from 30 to 350 m, of which the volcano Vekhne-Ingamakitskiy II has associated seismic activity at a depth of 15 to 20 km (Stupak, 1978a). Possibly included among the nine unnamed volcanoes in the IAVCEI (1973) catalog are the three late Quaternary volcanoes Dvoinoi, Yakutsk, and West Sakukan (Stupak, 1978a). The West Sakukan volcano is 30 m higher than the watershed and has an elliptical plan shape extending north-northwest–south-southeast, the long axis being 600 m and the short, 250 m. The cone rests on trachyandesite basalts of the middle portion of the plateau section. The outer zone of the cone comprises cinders and bombs and is succeeded inward by a lava elongated to the north. The lavas and bombs are nepheline-bearing analcime basalts and kulaite. The lavas contain xenoliths of pyroxene peridotite and granite gneiss. The Dvoinoi volcano is formed of two rounded peaks separated by 200 m and surrounded by a 0.1 km^2 lava field. The northeast peak is a 15-m-high lava dome, and the southwest peak is an eroded lava dome. The greater part of the lava of Dvoinoi is dense, black nepheline-bearing analcime basalt. The Yakutsk volcano is the oldest of the three. It has gently sloping sides, rises 30 m above the watershed, is oval in shape, and covers an area of 0.35 km^2. The lavas are hornblende alkaline basalts (kulaites), analcime kulaites, and analcime basalts.

In general terms, the early lavas tend to have high Na_2O (average, 6.44 weight percent) and K_2O (average, 3.22 weight percent content), and low CaO content (average, 5.17 weight percent), whereas the later flows are enriched in CaO (average, 9.91 weight percent) and depleted in SiO_2 (about 7 weight percent less than the early flows; see Appendix 7). The area is centered around 57°N, 118°E.

154. Chepe. This is a Holocene-age volcano with a summit crater. The volcano rises 380 m above its base and is composed of trachybasalt, trachyandesite, and andesite. The location is at 56.18°N, 117.55°E (IAVCEI, 1973).

155. Aku. This is a type of somma with a height of 220 m, composed of basalt, trachybasalt, and andesite. The volcano is of Holocene age and is located at 56.18°N, 117.45°E (IAVCEI, 1973).

156. Sini. A fissure type of eruption with present-day seismic activity located at a depth of 25 km below it (IAVCEI, 1973), Sini is presumably a central vent along the fissure, since it rises 250 m meters above its base. The volcano is composed of olivine basalt and is of Holocene age. The location is at 56.2°N, 117.35°E (IAVCEI, 1973).

157. Vitim Plateau. Lavas cover extensive areas of the Vitim Plateau and have been erupted from small cinder cones and larger volcanoes of the "central type." The lavas, of Neogene and Quaternary age, have thicknesses of up to 50 m. Some of the larger cones are Mushketova, Obrucheva, Lopatina, and Dombrovskiy (Bagin and others, 1971). These have basal diameters of about 1.5 km and heights of up to 150 m (IAVCEI, 1973). The lavas comprise a small volume of limburgites, some nepheline-bearing, and a large volume of alkali-olivine basalts (Kiselev and others, 1975). The location is centered around 54°N, 114°E.

158. Dgida (Dzhida) Basin. To the south and southwest of Lake Baykal lie a series of approximately east-west–trending basins containing extensive Quaternary and Tertiary basalts. Within the Dgida Basin there are basaltic alkalic and calc-alkalic flows and trachybasalts of Quaternary and postglacial age (Belov, 1958). There appear to have been at least two stages of eruption. There are lavas that overlie the 20-m terrace along the Dgida, Darkhintuy, and Khamney Rivers; along the Khamney and Khuraytsakir Rivers there are lavas that overlie the Dgida lavas, and in the Darkhintuy Valley the lavas are overlain by gravels, which are in turn overlain by flows from the Bartoy Valley. The Bartoy lavas are called shiklunites, which is presumably the same as shihlunite, the leucite basanite described from Wudalianchi (listing 20). An undefined number of cinder cones occur within the Dgida Basin, ranging in height from 55 to 240 m (IAVCEI, 1973).

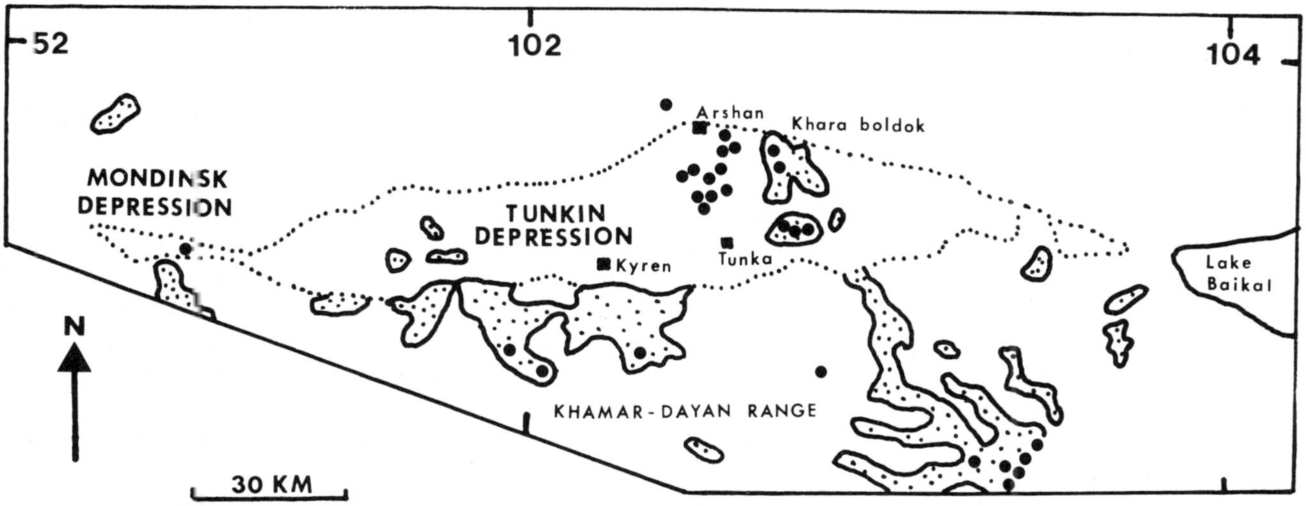

Figure 16. Distribution of basalts (stippled) and cones (dots) in Tunkin Depression (listing 160). Dotted line outlines the depressions. Filled squares are villages. (Modified after Florensov and Logatchev [1975].)

159. Khamar-Daban (Chamar Daban, Khamar-Dayan). The basalts of the Khamar Daban range from Oligocene to Quaternary in age. The "summit" basalts are mainly Miocene to Pliocene in age, whereas the "valley" basalts are Pliocene and Quaternary in age (Devyatkin and Smelov, 1979). The "summit" or plateau basalts once covered 3,000 km^2 and are from 100 to 350 m thick (Florensov and Logatchev, 1975). Along the Mongol-Daban River some 60 m of tuff underlie basalt flows (Belov, 1958). Figure 16 shows the disposition of some of the flows.

160. Tunkin Depression. A detailed stratigraphy of this basin has been obtained from drill cores. Between a depth of 1,000 and 2,100 m there are 65 basalt flows amounting to 600 m of thickness, and six tuffs of Miocene to Pliocene age amounting to 75 m of thickness. There are no volcanics between 500 and 1,000 m of depth. Within the top 500 m there are Quaternary deposits consisting of 12 basalt flows with a total thickness of 60 m, overlying 200 m of tuffs and sandy tuffs (Belov, 1958). Originally it was believed that there was a shield volcano near Talaya, but a later interpretation indicated that there were two cones, the remaining features being rafted fragments of destroyed cones. The rocks in this area include olivine basalt, palagonite basalt, and carbonatite basalt (Belov, 1958). In a more recent tour guide (Florensov and Logatchev, 1975), five groups of cones of Pleistocene to Holocene age were distinguished in and near the Tunkin Depression (see Fig. 16). The Talaya and Kuntinsky groups are situated over the zone of maximum sinking of the Tunkin Depression at its intersection with the Kyngarga fault. One of these volcanoes, "Holy Mountain" (Svyashchennaya Gora) is 30 to 40 m high and has a flat-floored crater 45 m in diameter. The Khobok group of slaggy cones, of which Khara-Boldok (Ulyaborskiy) is the highest and best preserved, rises 120 to 125 m and has an 85 to 90 m diameter summit crater. In the northern part of the Elovsky Spur is another group containing the destroyed Kovrizhka volcano. To the northeast of Kovrizhka lies Podgorny volcano, one of the largest in the Tunkin group. It is not known how these volcanoes relate to the Elovsky Spur volcanoes, they could be different names used by different authors for the same features.

161. Elovsky Spur. Between the Akhalik and Irkut Rivers are four largely destroyed cinder cones, composed mainly of scoria. These include Khobok in the northeast, Kuntinsk and Talovsk to the west, and Akhalik to the southeast (Belov, 1958). These could be alternate names for volcanoes of the Tunkin Depression.

162. Mondinsk Depression. This is a small east-west–trending basin just to the west of the Tunkin Depression. It contains a diatreme or explosion vent called Khulugayshi volcano. The olivine basaltic tuffs from the volcano overlie Quaternary gravels (Belov, 1958).

163. Oka Plateau. In the vicinity of the Dzon-Balyk River are the two cones, Kropotkin (52.6°N, 98.9°E) and Peretolchin (52.61°N, 98.93°E). These are both of Holocene age and composed of olivine basalt. The former is 90 m high; the latter, 120 m high (IAVCEI, 1973). The flows from these volcanoes extend 75 km down the river valleys (Belov, 1958). Another volcano, Strariy (52.6°N, 98.9°E) which is located nearby, is also composed of Holocene-age olivine basalt. Strariy is described as a cinder cone with a somma that rises 40 m (IAVCEI, 1973).

164. Orlik. Near the villages of Orlik (52.5°N, 98.8°E) and Khara-Buzhir are two isolated basaltic extrusions composed of scoria, lapilli, and bombs within an agglutinated mass on top of lava flows (Belov, 1958).

165. Northeast Tuva. The IAVCEI (1973) catalog listed two groups of volcanoes, one with seven members and the other with four, within the area bounded by 52.16° to 52.91°N, 97.83° to 99.33°E. The range in coordinates also includes the volcanoes of the Oka Plateau. Both groups are of Pleistocene age and are composed of olivine basalts and tuffs. The larger group is listed as a shield volcano with cinder and lava cones varying in height from 60 to 300 m, whereas the smaller group is called a cone-shaped volcano with a somma.

In the northeast Tuva region there is a great thickness of tuff overlain by the lower large flows of possible Oligocene age. The younger flows are of Quaternary age, predating the last glaciation (Belov, 1958). Gladkikh and others (1980) separated the region into two subareas: a north field where volcanic rocks occur on stream divides as isolated areas hundreds of meters wide and up to 8 km long, and a southern field of small individual areas. There are no more than seven flows, ranging in thickness from 4 to 30 m.

166. Unnamed. A Quaternary olivine basalt volcano group of fissure eruptions are located around 51°N, 96° to 97°E (IAVCEI, 1973).

167. Khamsary-Bol'shoy Yenisey. Overlying a 100-m-thick succession of the lower (Oligocene?) lavas is a sequence of tuffs 200 to 350 m thick. The eroded surface of the tuffs is covered by younger flows. Both the tuffs and the younger flows have been interpreted to be of Tertiary age (Belov, 1958).

168. Upper Bol'shoy Yenisey. The IAVCEI (1973) catalog listed two volcanoes in this area; the southernmost is the Pleistocene shield volcano, Derbi-Tayga (52.2°N, 98°E), and the other is the Holocene cinder cone, Ulag Arginsky (52.33°N, 98°E). Derbi-Tayga is composed of olivine basalt and rises 700 m, whereas Ulag Arginsky is composed of basalt.

169. Other volcanoes in the Sayan Baykal Area. Belov (1958) also recorded the following: plateau basalts overlying a tuff with a Miocene fauna along the Tissa River, a 0.4-m layer of tuff under a basalt along the Taitorka River near the town of Shimka, a Tertiary central volcano on the Chika river (Bol'shoy Kamya), a shihlunite-stratified cone in the Barun-Khobol Basin, and pyroclastic deposits between the Ii and Uda Rivers. He also noted that the East Sayan highlands contain 26 flows that go to form the Arlykgol'sk field and an additional 21 flows that form the Saganarsk field. (These widely dispersed localities are not indicated in Figure 1b.)

PART II

VOLCANISM, TECTONICS, AND TIME

The area encompassed by this survey is approximately 21.7 by 10^6 km^2, nearly three times the size of North America. Given the paucity of information available on this vast area, it is only possible to make broad generalizations about the volcanic stratigraphy.

STYLE VARIATION WITH TIME AND LOCATION

There are few large composite volcanoes in mainland Asia; P'aektu-san on the China-Korea border is the best known of those that have recently been active. Other large composite volcanoes include Mt. Popa, Mt. Loimye, and Taungthalon in Burma; possibly Dayingshan, Yaowushan, Luzongshan, and Peishan in China; the early Cenozoic calderas of the Okhotsk-Chukotki belt of the Soviet Union; and some calderas in northern Tibet. A few silicic domes are found, but the large majority of the volcanoes are plateau basalts, lava shields, and cinder cones. A few of the larger cones, such as Bilibina and Balagan-Tas, might be considered as small composite cones. A listing of the various types of volcanic features, along with their locality numbers, is presented in Appendix 4. Figure 17 is a plot of height versus frequency for the Asian volcanoes that clearly shows the lack of large volcanoes, especially when allowance is made for those volcanoes for which no dimensional data are available. Other dimensional parameters have been plotted in Figure 18, where they are compared with empirically derived curves compiled by Wood (1978, 1980). The scatter in these diagrams arises largely from the varying degrees of erosion of the volcanoes.

It is difficult to establish the variation of eruption style with time, as many of the volcanoes are undated and erosion and burial obscure the early structures. For example, a field of cinder cones could be buried by a plateau fissure eruption or a large shield formed early in the Tertiary could be eroded to its roots. Bearing in mind these difficulties, a broad overview of the style-time variation will be presented.

The eastern seaboard of Asia was subjected to extensive eruptions of andesitic and rhyolitic materials from Mesozoic through Pliocene time, often in the form of ignimbrite eruptions. This zone extends from the Arctic Sea in the north, down and along the western side of the Sea of Okhotsk, through Sikhote-Alin, northeast China, Korea, and into southern China. The Late Mesozoic–Early Cenozoic evolution of this region has recently been reviewed by Takahashi (1983). The volcanic eruptions within this area became less siliceous with time so that, in the majority of cases, large shields or plateau basalts constitute the most recent extensive eruption phase. Often these basaltic plateaus are dotted with cinder cones or small shield volcanoes. Small, highly silicic domes have been found in the northern Sikhote-Alin area. P'aektu-san differs from the other volcanoes of the area in that the basaltic plateau has been surmounted by a large, pantelleritic and comenditic strato-volcano.

In central north Manchuria there are large lava fields and associated cinder cones of Holocene age that appear to be in areas not subjected to volcanism in the earlier Cenozoic. In western Manchuria there are scattered basalt plateaus and lava fields in an area that suffered extensive Jurassic-Cretaceous volcanism. These lavas form the apparently Oligocene-age Mongolian plateau basalts and are overlain by Pleistocene lavas and cones associated with fissures. The early plateau basalts have also been recovered from drill holes in downdropped, fault-bounded basins (Zhou and Armstrong, 1982). A fissure eruption style is also found in the early flows of the Anyuy group, the alkaline flows of northeastern Burma, Sini, the east Sayan highlands, and Indochina.

The Thailand eruptives are often highly altered and there are few references to volcanic edifices, which suggests that many may have been fissure-fed. Some eroded necks may, however, have originally been central vent volcanoes. The eruptions in the remainder of Indochina were also fissure-fed, producing vast plateau basalts, apparently commencing during the Miocene but having a peak output near the Pliocene-Pleistocene boundary.

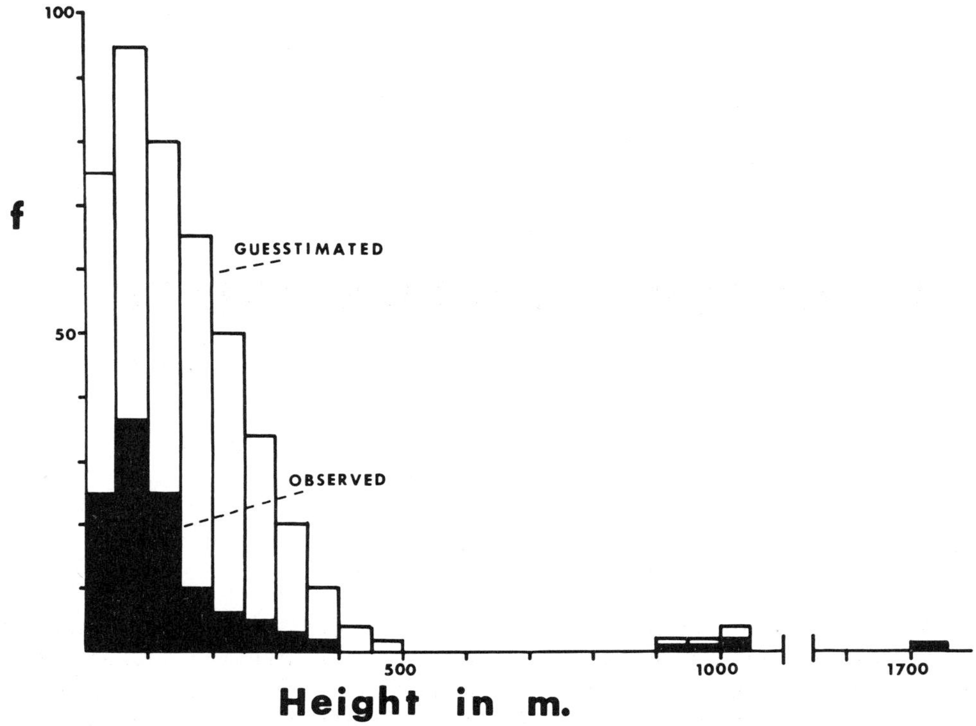

Figure 17. Histogram of height/frequency distribution of Asian volcanoes described in text. Curve labeled "OBSERVED" is based on data presented in Appendix 6. There is little numeric data for many of the volcanoes; therefore "guesstimated" values have been interpreted, based on descriptions of approximate numbers and types of volcanic features present at different localities. Distribution reflects fact that vast majority of the volcanoes are cinder cones, fissure eruptions, and maars.

With time, volcanism became more explosive; lava cones, cinder cones, and maars were formed within the plateaus. The Indochina volcanics are notable for their frequent association with gem-quality zircons and sapphires.

The Burmese volcanics are aligned in two approximately north-south–trending belts, or three belts if the offshore islands are included (Stephenson and Marshall, 1984). The Late Cretaceous–early Tertiary igneous rocks are mainly ultrabasic, and include lherzolites, gabbros, dunites, and peridotites, probably formed on the ocean floor. Also of this age are the basaltic tuffs and andesites of the Mt. Loimye area. In the Pliocene and Pleistocene a great variety of rock types were erupted. At Mt. Popa there are early silicic lavas overlain by a basaltic plateau and capped by a basalt and basaltic andesite composite cone. The environment of deposition of the Shwebo district pillow lavas was probably fluvial or lacustrine, as they are intercalated with sediments of the terrestrial Lower Irrawaddy Group (Bender, 1983). The nearby Quaternary lavas of the Singu Plateau were erupted from fissures associated with the Sagaing fault system. The explosion craters of the Chindwin area are of a type found in few other places in eastern Asia and possibly owe their form to groundwater associated with the Chindwin river. The phreatic craters around Chaun Bay in the Soviet Union may also be of this type.

In the Tengchong area of southwest China, the Tertiary lavas occur as flood basalts with a total area of 490 km^2; the more siliceous Quaternary eruptions are from central vents and cover about 265 km^2 (Zhu and others, 1983). The area is one of current hydrothermal activity.

Information concerning the volcanics of western and central China has been largely derived from LANDSAT imagery. Most volcanoes appear to be cinder cones, composite, or calderas. Some structures of the Bamaoqionzong area have been described (Deng Wanming, 1978a) as edifices dissected into table mountains with columnar upper portions. The high elevation of this area would not rule out the possibility that these features could be tuyas (Allen, 180), although there was no mention of pillow structures.

The Tertiary and Quaternary eruptions of northern Mongolia appear to be largely plateau basalts. Lavas are also found in the depressions associated with the fold belts that wrap around the southern portion of the Siberian Platform. The same appears to hold for the East Sayan region, but the Vitim and Stanovoy areas appear to be late plateau eruptions with cinder cones, associated with the Okhotsk-Baykal zone.

In central Mongolia, the majority of the most recent eruptions appear to have infilled river valleys, although plateau basalt remnants of Oligocene and Miocene age occur in the central part of the country. The peculiar structures described by Tikhonov

a

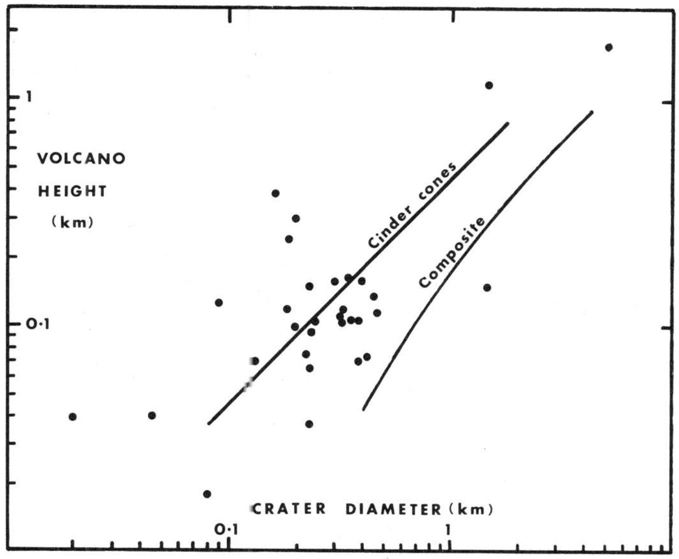

b

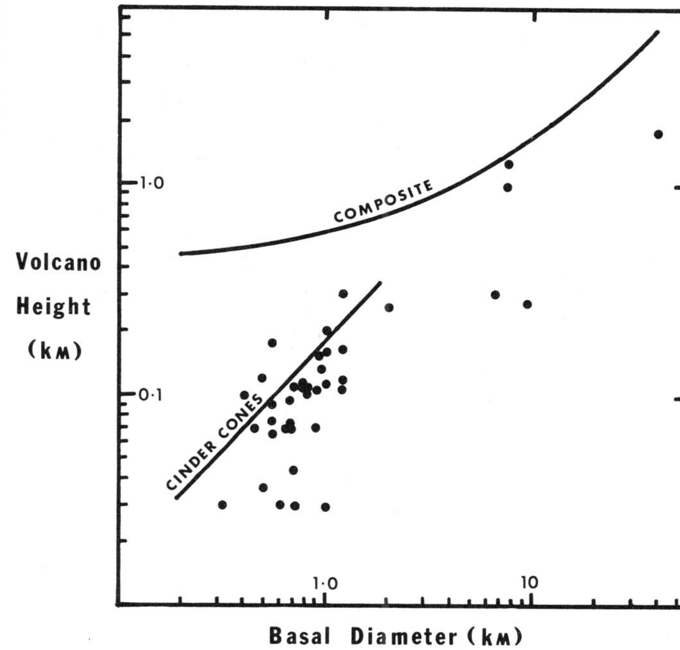

Figure 18. a, Volcano height/crater diameter plot of cataloged volcanoes. Deviation of values from curves for fresh cinder cones and composite volcanoes (from Wood, 1978, 1980) reflects erosional modification. b, Volcano height/basal diameter plot for cataloged volcanoes. Curves for cinder and composite cones from Wood (1978, 1980). c, Crater depth/crater diameter plot for cataloged volcanoes. Limited range is reflection of paucity of available data.

c

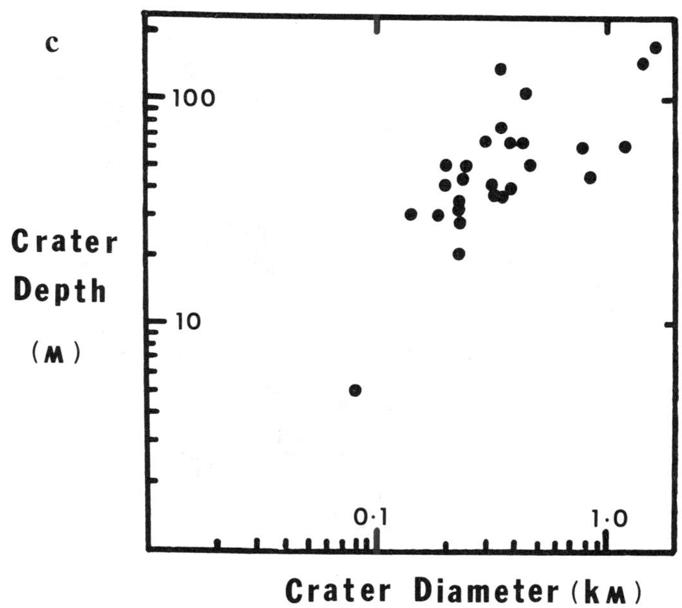

(1977) in southeast Mongolia (listing 109) appear to have many of the characteristics of central-peak impact basins and their mode of origin is suspect.

In summary, the eastern coast of the Soviet Union and the western volcanic zone of Burma were marked by early andesitic and silicic volcanism, followed by the development of large basaltic plateaus. Conversely, at P'aektu-san the early volcanics were basaltic, producing a plateau on which a silicic cone has been constructed. In Indochina and interior Asia the commonest expression of volcanism is the basaltic plateau, often produced from fissures, commonly associated with graben structures. Activity commonly terminated on the plateaus with the formation of cinder cones and maars.

VOLUME CHANGE WITH TIME AND LOCATION

In location where there are extended vertical sections, it is apparent that, in most cases, volcanism has been episodic rather than continuous; periods of extensive activity have been separated by even longer periods of inactivity. Few areas have been continually active throughout the Cenozoic. Areas that have been the site of volcanic activity for considerable periods include Sikhote-Alin (Cretaceous to Recent), the Jade Mines district (Cretaceous to Quaternary), the Udokan Plateau (Miocene to Quaternary), the Vitim Plateau (Miocene to Quaternary), the Khamar-Daban (Oligocene to Quaternary), and the Tunkin Depression (Miocene to Quaternary). A list of the ages of volcanism at the various localities described in Part 1 of the text is presented in Appendix 4.

The most categoric statement that can be made about Asian volcanoes is that they are not as active now as they were in Oligocene-Miocene and Pleistocene time. Only 13 volcanoes have been found that may have erupted within the last few millenia, and few of these are verified observations. The *Catalog of Active Volcanoes of the World* (IAVCEI, 1951–1975) lists only four historically active volcanoes: Laoheishan (1720), Huoshaoshan (1720), an unnamed volcano in Tibet (1951), and Balagan-Tas (Indighirsky?) (1770s). This compilation also adds another

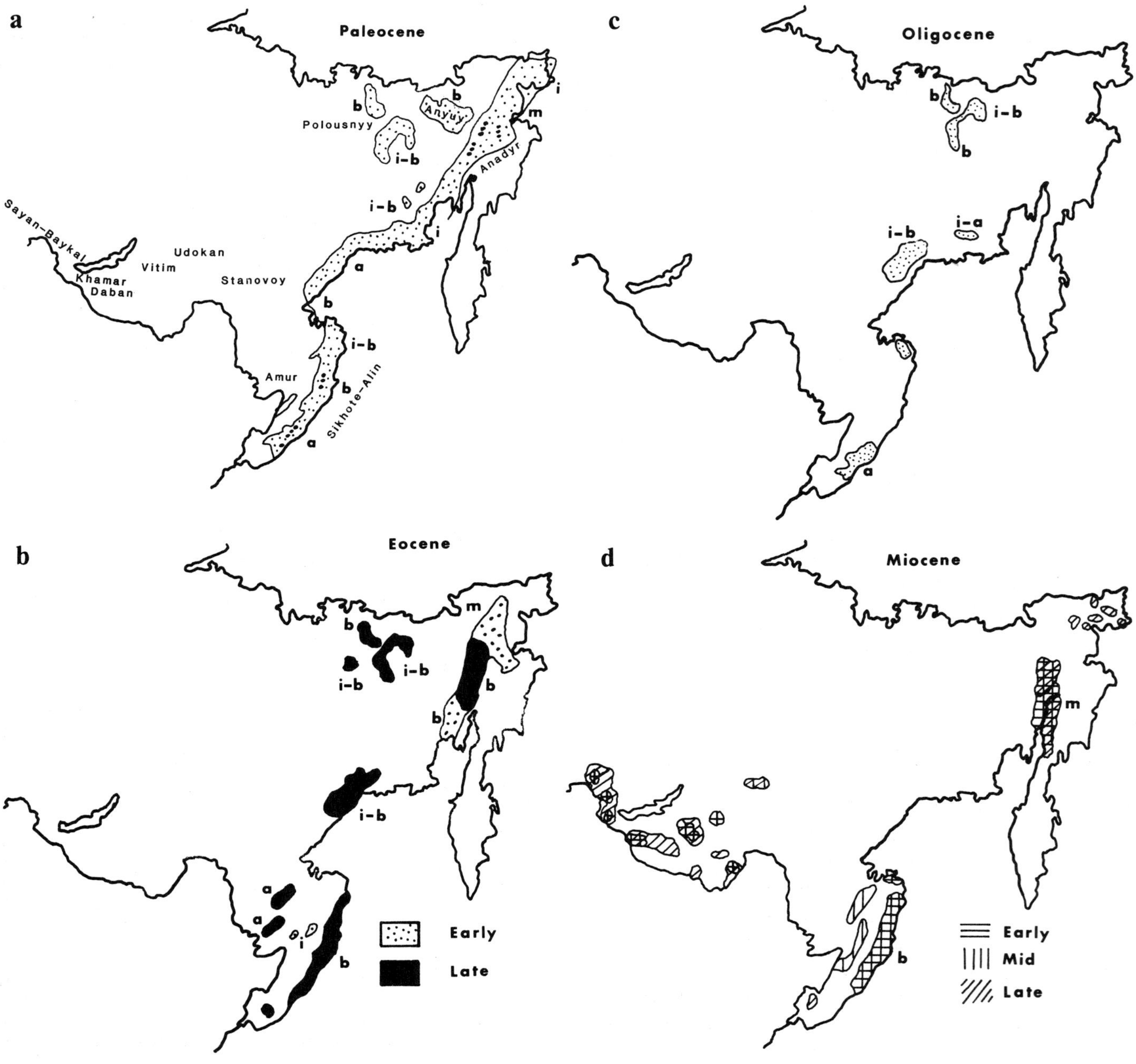

volcano in Tibet (1973), P'aektu-san (1597 and 1702?), Dayingshan (1609?), Peishan (first and seventh centuries), Anyuy (14th and 18th centuries), Mount Popa (442 BC), Aluchin (1000 AD), possibly some cones in the Dariganga region, and even less possibly, Turfan and Ou-roumtsi. The volume of material issued by these eruptions is insignificant in comparison with individual eruptions of the earlier plateau basalts.

The largest outpourings of volcanic materials appear to have taken place during the Oligocene and Miocene periods to form the plateau basalts. However, there are poor dimensional and time constraints on these eruptions. For example, the plateau basalts of the Khamar-Daban range are described as covering an area of 3,000 km^2, but estimates of its thickness range from less than 150 to 350 m. Similarly, although the basalt remnants in the area to the west of Beijing have been described as Oligocene in age and 150 m thick, I could find no estimates of its original areal extension. There also appears to have been a tendency, at least among the early geologists in the area, to assign an age to a particular plateau basalt because it appeared to be morphologically similar to a plateau basalt elsewhere that had been dated. The apparent narrow time span of the early plateau basalts may therefore be somewhat artificial. Carbonnel and others (1972) recognized that the southern Indochina plateau basalts were erupted in three well-defined periods at approximately 0.64, 1.3,

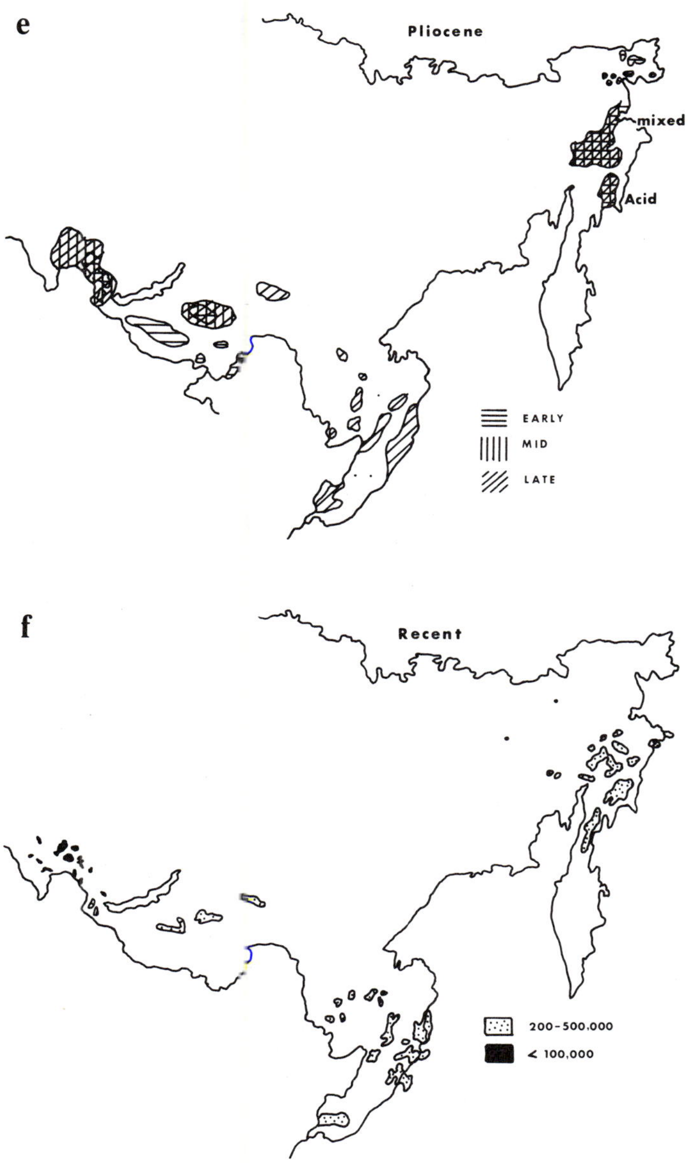

Figure 19a-f. Series of paleogeographic maps of eastern Soviet Union, showing distribution of volcanic materials through the Cenozoic. a = acidic, b = basaltic, i = intermediate, m = mixed compositional types. Rock types have not been included on diagrams illustrating Miocene through Recent distribution because separation based on both rock type and age would have resulted in unreadable figures. (Based on Vinogradov [1967].)

and 2.4 m.y. ago. More recently acquired data (Barr and Macdonald, 1981) has extended the time range of these eruptions from approximately 12.5 to 0.45 m.y.

Appendix 4 lists the ages of the volcanic complexes for which there is available information. The best database is for Indochina, recently summarized by Barr and Macdonald (1981). Cinder cones predominate among the most recent eruptions, probably solely as a result of their propensity to erode and become buried with time, as opposed to their not being erupted in earlier periods. The plateau basalts are largely of late Eocene to Miocene age, and extensive basaltic lavas were produced during the Late Pliocene to Pleistocene. This is a gross oversimplification, since the data is sparse for much of the area. In areas in which more detailed observations have been made, for example the Soviet Union, a complex variation in time and location of the erupted materials is evident. Paleogeographic maps of the Soviet Union (Fig. 19) illustrate that during the Paleocene there was a volcanic belt of largely acidic to intermediate materials extending down the whole of the eastern side of the Soviet Union and occupying a few areas in the northeast. These belts contracted during the Eocene, but acidic activity became more extensive in the Amur region (listings 142 through 144). A contraction in the area of volcanism continued through the Oligocene epoch. During the Miocene, activity commenced in the East Sayan (listing 165)–Baykal (listing 158)–Vitim (listing 157)–Stanovoy (listings 153 through 156) areas, while basic materials were erupted in the Sikhote-Alin (listings 137 through 152) and Amur areas. The Pliocene marked the most extensive development, in terms of area, of the East Sayan–Stanovoy belt, while widely separated volcanic centers were in eruption on the east coast. By 200,000 B.P., activity had largely ceased in the Sikhote-Alin and Vitim districts, and by 80,000 B.P. large volcanic centers were restricted to the East Sayan region. In the last 10,000 years there have been small isolated eruptions in the northeast Soviet Union and the East Sayan region, and extensive volcanism on the Kamchatka Peninsula which has been neglected in the present study.

CHEMISTRY VARIATION WITH TIME AND LOCATION

The geochemistry of the Asian volcanics region, which was recently reviewed (Whitford-Stark, 1983), has been subdivided into 14 geochemical provinces (Fig. 20). Thus this section of the present paper is a brief review and update of the material in the earlier article. Geochemical analyses for samples from individual locations are presented in Appendix 7. Table 1 cross-references the geochemical provinces illustrated in Figure 20 with the localities from Part 1.

Province A. This province extends from the Chukotki Peninsula in the northeast, down along the west side of the Okhotsk Sea and into Sikhote-Alin. As might be expected for a region 4,500 km in length, the erupted materials exhibit extensive spatial and temporal geochemical variation. In the southern part of the province, Bersenev and others (1966) separated four episodes of Cenozoic volcanic activity in the southern Sikhote-Alin: a Paleocene episode of equal proportions of acidic and intermediate rocks, an Eocene-Oligocene period of thin acid and basic flows, a Miocene episode of mainly basaltic material with little acidic, and a Pliocene to Recent period of entirely basaltic material. Farther north, Cenozoic plateau basalts overlie some 800 m of Cretaceous volcanics in the north Okhotsk region (Speranskaya, 1966).

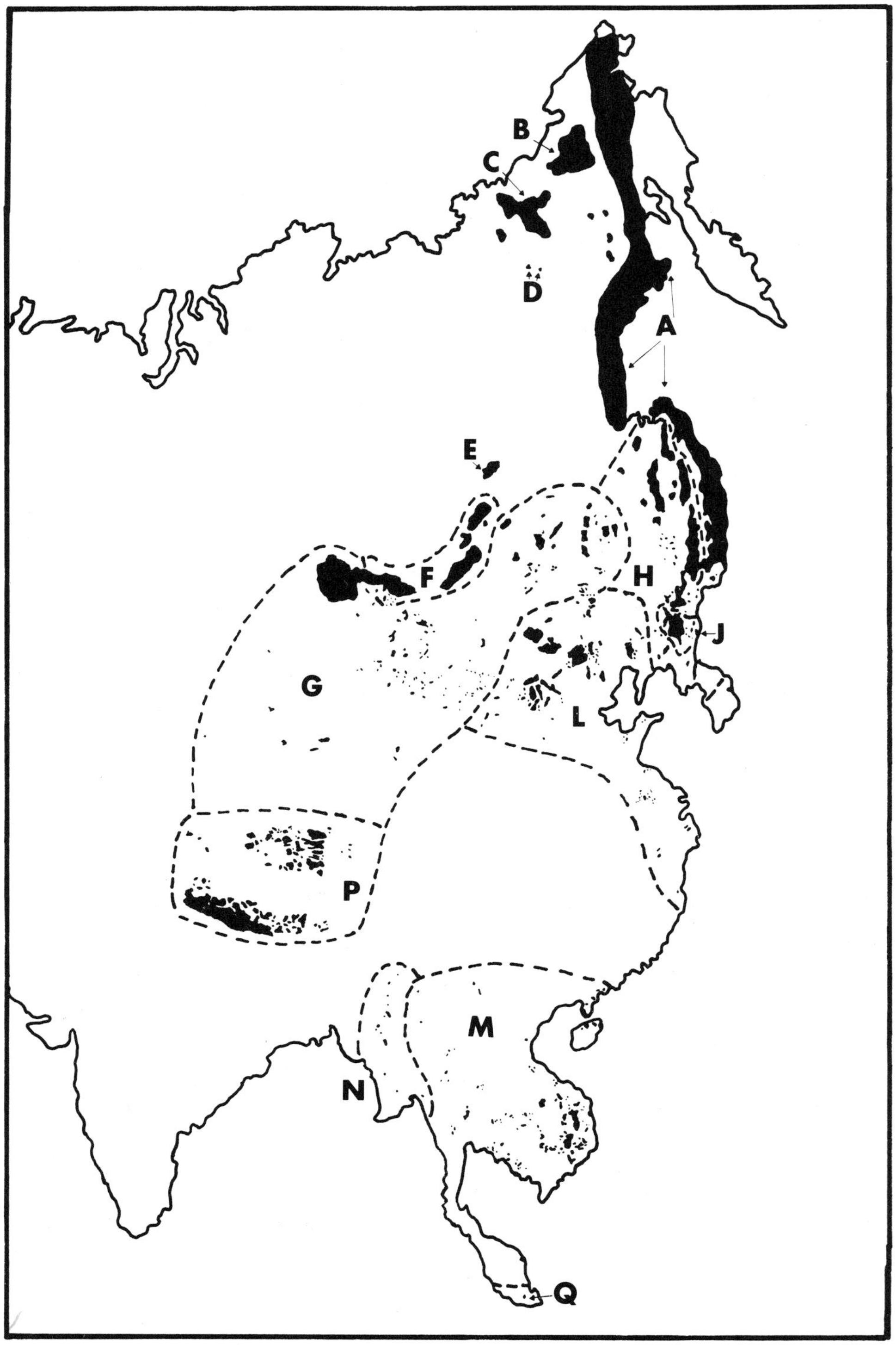

Figure 20. Geochemical provinces (A-Q), outlined in dashed lines, of mainland Asia exclusive of Kamchatka and Koryak Peninsulas. These provinces provide the framework for the geochemical and tectonic evolution of mainland Asia outlined in Part 2 of the text. Areas outlined in black for region corresponding to Figure 1a represent areas in which volcanic rocks currently crop out. Similarly patterned areas in region corresponding to Figure 1b represent volcanic deposits defined from paleogeographic reconstructions. Rocks within each province have geochemical or temporal affinities that are distinct from those of adjacent provinces; subprovinces could be defined (e.g., Barr and Macdonald, 1981) if more detailed information were available. (From Whitford-Stark [1983]; reproduced by permission of Elsevier.)

Likewise, Cretaceous ignimbrites in the Anadyr Trough (126) were overlain by Paleogene basaltic shield volcanoes (Zhuphakin and others, 1977). The most recent eruptions within the province have been basaltic, for the most part, and apparently subalkalic (Vinogradov, 1967; Plakhotnik, 1968).

Province B. The Aniuski Range (128) in northeastern Siberia contains Paleocene basic lavas (Vinogradov, 1967) and Pleistocene to Holocene hawaiites, benmoreites, potassic basalts, and trachybasalts (Chasovitin, 1963; Gorodinskiy and others, 1967; Ustiev, 1959a, b).

Province C. Intermediate and basic volcanics were erupted in this province from the Paleocene through the Oligocene (Vinogradov, 1967). Farther west, early trachytes, liparites, and dacites are found in the Polousnyy Mountains (134) and the Ulakan Cis Range, and Lower Pleistocene hawaiites are found along the Indighirka River (Ustiev, 1974).

Province D. In eastern Siberia, a Quaternary liparite forms the Majak dome (listing 133) (IAVCEI, 1973), and late Holocene hawaiites were erupted from Balagan Tas (listing 132) (Argunov and Gavrikov, 1960; Gorshkov, 1970).

Province E. The Udokan Plateau (listing 153) witnessed four episodes of volcanism (Stupak, 1978b): a Neogene episode of fissure eruptions of basaltic trachybasalt, trachyandesitic basalt, trachyandesite, and trachyte; an early Pleistocene series of fissure eruptions of trachybasalt and trachydolerite; a Pleistocene episode of central vent eruptions of analcime basalt and analcime kulaite (see Appendix 2 for rock definitions); and Holocene central vent eruptions of alkali trachyte. The plateau contains some of the most sodic and silicic rocks of the Baykal-Okhotsk zone; some of the rocks have pantelleritic affinities (Stupak, 1978a).

Province F. The volcanism of the Baykal rift zone has been summarized by Kiselev and others (1978), who found a Late Cretaceous to Early Paleocene episode of basaltic pyroclastics and lavas in the Tunkin depression (listing 160), Miocene to Early Pliocene fissure eruptions of alkaline and subalkaline basalts in the Vitim Plateau (listing 157), Khubsughol (listing 117), Sayan (listing 158F), Khamar-Daban (listing 159), and Tunkin depression areas, Early Pleistocene fissure eruptions of alkaline and subalkaline basalts in the same areas, and, finally, in the Pleistocene to Holocene, eruptions of alkali olivine basalts and basanites. Tholeiites are largely confined to the Tunkin depression and the Kitoy area (listing 161). Within individual areas there is a trend of increasing alkalinity and basicity with decreasing age; overall there appears to be a progressive increase in alkalinity from west to east. In many areas the younger lavas are predominantly hawaiites, sometimes accompanied by shihlunites (see Appendix 2 for rock definition) (Belov, 1958).

Province G. This large province, consisting of northern China, Mongolia, and part of the Soviet Union, is composed almost exclusively of alkaline volcanics. In northern Manchuria, Pliocene lavas consist of olivine leucitites and tephritic leucitites; Pleistocene lavas consist of leucite trachytes, phonolites, and tephritic phonolites; and Holocene lavas of similar composition to the Pleistocene lavas (Ogura, 1969; Tomita, 1969; Tsung-pu, 1960; Zheng Xiangshen, 1983). Devyatkin and Smelov (1979) recognized six episodes of basaltic volcanism within Mongolia: Eocene, Oligocene, Miocene, Pliocene, Lower Pleistocene, and Pleistocene-Holocene. The largely Miocene volcanics of the Khangay (listing 123) are limburgites, analcime andesite-basalts, and trachybasalts. The Pleistocene-Holocene rocks of the Dariganga area (listing 125) consist of picritic basalts, limburgites, analcime, andesine, and oligoclase basalts, and trachybasalts. The lavas of the Gobi Altay (listing 124) are predominantly Eocene to Oligocene in age and are preserved as dikes, necks, and layers of basalt. These consist of limburgites, augitites, analcime basalts, trachybasalts, and trachyandesites. Devyatkin and Smelov (1979) proposed an increasing depth of origin of the basalts from the earliest lavas to the youngest. The rocks of central Mongolia are almost exclusively trachybasalts; exceptions include shihlunites, leucite tephrites, and basanites in the Taryatu Depression (listing 122) (Kepezhinskas and others, 1975), phonolitic trachybasalts in the Khangay region (listing 123) (Filippov and others, 1976), and leucite basalts along the Bulgan fault zone near the Gobi Altai (listing 124) (Kepezhinskas and others, 1975). The extreme northwest of the province, which is an extension of the Baykal rift zone, includes the basalts of northeast Tuva, which are intermediate between alkali olivine basalts and continental trachybasalts (Gladkikh and others, 1980).

Province H. This province follows the northeastern coast of China and extends into the Soviet Union. The volcanics take the form of lavas and plateaus of middle to upper Cenozoic age and comprise trachybasalts, K-alkalic basalts, and nephelinites (Kuno, 1953; Ogura, 1969; Tomita, 1969). Leucite basalts have been recorded from western Sikhote-Alin (IAVCEI, 1973).

Province J. Located close to the China-Korea border, this province is unusual in that it is the only province that contains Cenozoic volcanics with pronounced peralkaline affinities. The volcano P'aektu-san (listing 5) is underlain by a Pliocene hawaiite basalt plateau, which is overlain by a Pliocene shield of alkaline trachytes, comendites, comenditic trachytes, and pantellerites. Pleistocene eruptions consisted of long flows of pantellerites with

TABLE 1. LOCALITY NUMBERS FROM PART 1, CROSS-REFERENCED TO GEOCHEMICAL PROVINCES OUTLINED IN FIGURE 20

Geochemical Province	Locality Numbers
A	126, 136, 137-141, 152
B	127-131
C	134
D	132, 133
E	153-156
F	117, 157-163
G	18-27, 51-53, 109-116, 118-125, 164-169
H	1-3, 6-9, 11-17, 56, 142-151
J	4, 5
L	10, 28-45
M	46-49, 65, 69-71, 73-108
N	57-68, 170
P	50, 54, 55
Q	72

tholeiites, hawaiites, and K-alkalic basalts (Tomita, 1969; Ogura, 1969; Cong Bolin and others, 1979). In the nearby Kilchu-Myongch'on district of Korea, early Miocene eruptions consisted of ne-normative analcite basalts and olivine trachybasalts. These were followed in the Pliocene by basalt eruptions, followed in turn by two sequences of trachytes, rhyolites, and comendites. A sequence of basalt through trachyte was repeated twice during the Pleistocene (Tomita, 1969). The younger lavas in this area were quartz normative. Overall, the Kilchu volcanics tend to be more potassic than those of P'aektu-san.

Province L. The volcanic rocks of this part of central eastern China have been the subject of several geochemical syntheses (e.g., Tsung-pu, 1960; Cong Bolin and others, 1979; Zhou and Armstrong, 1982). The earliest eruptions appear to have been of Upper Eocene hawaiites in Fuxin (listing 40) and K-alkalic basalts along the Xialiaohe (Cong Bolin and others, 1979). Late Oligocene tholeiites, Na-alkalic basalts, and hawaiites form the plateau basalts of Hanruopa. Hawaiites and Na-alkalic basalts of Miocene age occur in Weizhang (listing 30) (Lee, 1939). A trachybasalt has also been reported from this area (Tsung-pu, 1960). Other early Tertiary basalts from the province are predominantly hawaiites and Na-alkalic basalts, although some nephelinites and tholeiites have been reported from the Nanjing area (listing 43) (Cong Bolin and others, 1979). In eastern China, Zhou and Armstrong (1982) recognized two types of early Tertiary basalts: the Beijing-type, which is usually hypersthene and quartz normative, and the Bohai-type, which is predominantly hypersthene and olivine normative.

Through the late Tertiary and Quaternary, hawaiites, nephelinites, and sodic alkalic basalts continued to be erupted although trachybasalts became relatively more abundant (Cong Bolin and others, 1979). Late Tertiary and Quaternary basalts are overwhelmingly alkalic to very alkalic. In general, within this province the basalts become increasingly alkalic with decreasing age and with increasing distance westward from the coast (Xie Guanghong, 1981). The southern part of Korea was characterized by the eruption of olivine basalts and andesite during the Miocene and then volcanic activity ceased. These basalts have affinities with those of the Japanese volcanic province (Tomita, 1935). Central Korea experienced voluminous eruptions of alkali olivine basalts during the Pleistocene.

Province M. Zhu and others (1983) defined five chemical groups of volcanics in the Tengchong district: Pliocene high-Al basalt; Late Pliocene high-Mg basalt; Early Pleistocene high-K olivine basalt; Late Pleistocene high-Al andesitic basalt and Early Holocene high-Al pyroxene-olivine basalt; and Late Holocene high-K andesitic basalt.

The geochemistry of Indochina has been recently summarized by Barr and Macdonald (1981), who found that the oldest basalts are generally tholeiites or tholeiites transitional to hawaiites, whereas the younger basalts are increasingly alkalic. The majority of the province is composed of hawaiites of Pliocene-Pleistocene age. The basanitoid rocks are volumetrically small, forming plunge, isolated vents, or small lava flows capping older, more voluminous, less undersaturated flows. Holocene eruptions have taken place on submarine volcanoes located close to the mainland coastline (Simkin and others, 1981).

Province N. Stephenson and Marshall (1984) illustrated a tripartite division of the Burmese volcanics from a tholeiitic–calc-alkalic composition of the offshore islands, through a calc-alkaline–alkaline group forming the central belt, to an alkaline field in the eastern belt. The majority of the volcanics are of Late Pliocene to Pleistocene age, although Mount Popa (listing 66) may have erupted in 442 BC.

Province P. Tibet is characterized by predominantly silicic and potassic volcanics with leucitites, leucite phonolites, tristanites, icelandites, dacites, and rhyolites. Deng Wanming (1978) defined two Quaternary geochemical provinces, the strongly alkaline lavas of the Tibetan block represented by Bamaoqionzong (listing 50) and the calc-alkaline sequence of the block margin represented by Yongbohu (listing 50). The calc-alkaline silicic volcanics are, however, more alkaline than the calc-alkaline rocks of Province N (Whitford-Stark, 1983). The evolution of Tibetan volcanics from the Triassic to the present has been briefly reviewed by Jin Chengwei (1981).

Province Q. The fissure-fed Kuantan (listing 71) lavas consist of nepheline basanites and olivine basalts dated at 1.6 m.y. (Barr and Macdonald, 1978). The nearby lavas of Segamat (listing 72) are ultrapotassic, consisting of leucite basanites, leucite tephrites, and potassic ankaramites. These have recently been dated as at least Paleocene in age (Barr and Macdonald, 1981) and are therefore probably unrelated to the Kuantan lavas.

SUMMARY

In overall terms, two distinct Cenozoic volcanic environments can be recognized on mainland Asia: a calc-alkaline sequence associated with the near-coastal regions of Burma, China, the Soviet Union, and the Tibetan plateau, and a progressively more alkaline sequence with increasing distance inland (see Figs. 21, 22). On top of this latter trend is superimposed a trend of increasing alkalinity with decreasing age at individual localities. The predominantly calc-alkaline and alkaline nature of the rocks is well illustrated by AFM diagrams (Fig. 22), which show a complete lack of an iron-enrichment trend.

TECTONIC SETTING

The most conspicuous feature of Figure 1 is the almost total absence of volcanic materials in south-central China and in the Soviet Union inland of the coastal belt and north of the Sayan-Okhotsk belt. This distribution appears to be real and not a function of a lack of data. The areas correspond, respectively, to the stable blocks of the Yangtse craton and the Siberian Platform. Two models, not entirely dissimilar, have been invoked to explain the recent tectonic evolution of Asia. Tapponnier and Molnar (1977) and Tapponnier and others (1982) treated the entire area as a single block, across which the stresses resulting from the

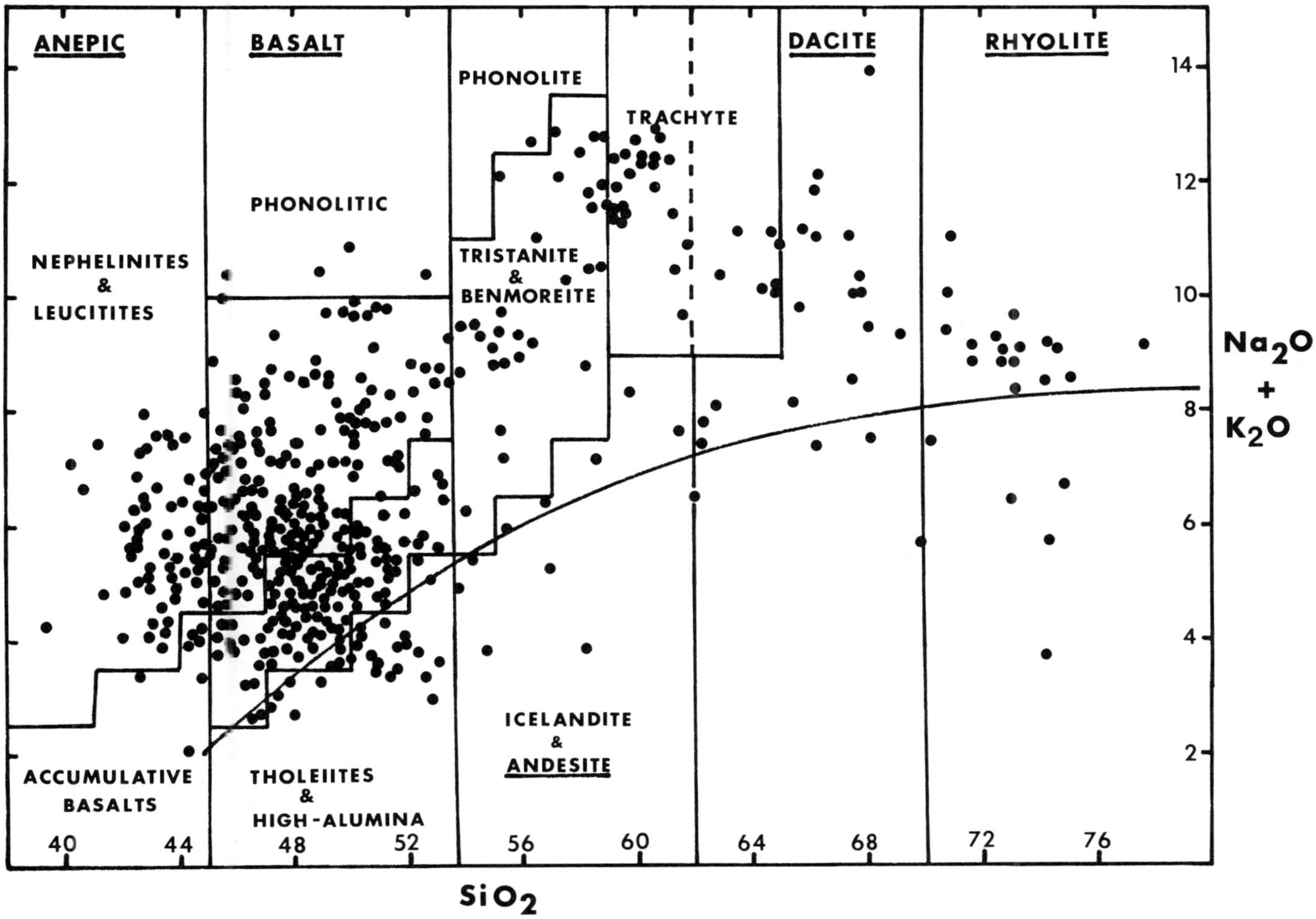

Figure 21. Plot of $Na_2O + K_2O$ versus SiO_2 for samples listed in Appendices. Also shown is geochemical classification scheme of Middlemost (1972) and curve of Miyashiro (1978) separating subalkalic from alkalic rocks. Principal rock "clans" are underlined. ANEPIC is acronym for ankaramite-nephelinite-picrite. Note preponderance of alkaline rocks. (From Whitford-Stark [1983]; reproduced by permission of Elsevier.)

collision of the Indian block are transmitted. Krasnyy (1967) viewed Asia as being composed of stable blocks bounded by mobile belts where most of the tectonic activity was taking place. Zonenshayn and Savostin (1981) made a similar but plate tectonic interpretation, treating the tectonic events of Asia in terms of the interactions between a series of plates and microplates. The present discussion attempts to outline the relationships between volcanism and tectonics in Asia. Figure 23 locates some of the major tectonic features of the Asian continent. Figure 24 summarizes the eruptive, geochemical, and tectonic evolution of the region.

Province A. In the eastern Soviet Union there was a decline in volcanism accompanied by a change from acidic to basaltic volcanism during the Eocene. This was matched by the establishment of volcanism on the Kamchatka Peninsula by the Oligocene and reflects the eastward shift of the subduction zone away from its previous location defined by Province A in Figure 20 and known as the "Evenian belt" (Nalivkin, 1973). This change appears to have been earliest in the north portion of the belt and latest in the Sikhote-Alin region east of the Ussuri River suture. Parfenov and Natal'in (1979) have attributed the rapid decline of volcanism to the wedging-in or suturing of the Sea of Okhotsk microcontinental block at the end of Cretaceous to Early Paleogene. The later termination of volcanism to the south can be ascribed to the progressive formation of the Sea of Okhotsk back-arc basin during the Eocene (Watanabe and others, 1977) and the Japan and Yamoto basins no earlier than the mid-Eocene and early Miocene, respectively (Sillitoe, 1977). The formation of the back-arc basins effectively moved the active subduction zone eastward so it was no longer directly beneath the mainland Asian plate.

Province B. Paleocene volcanism in the Aniuski range was possibly back-arc to the Evenian belt. The Pleistocene and Holocene eruptions were confined to fault-bounded valleys (Ustiev,

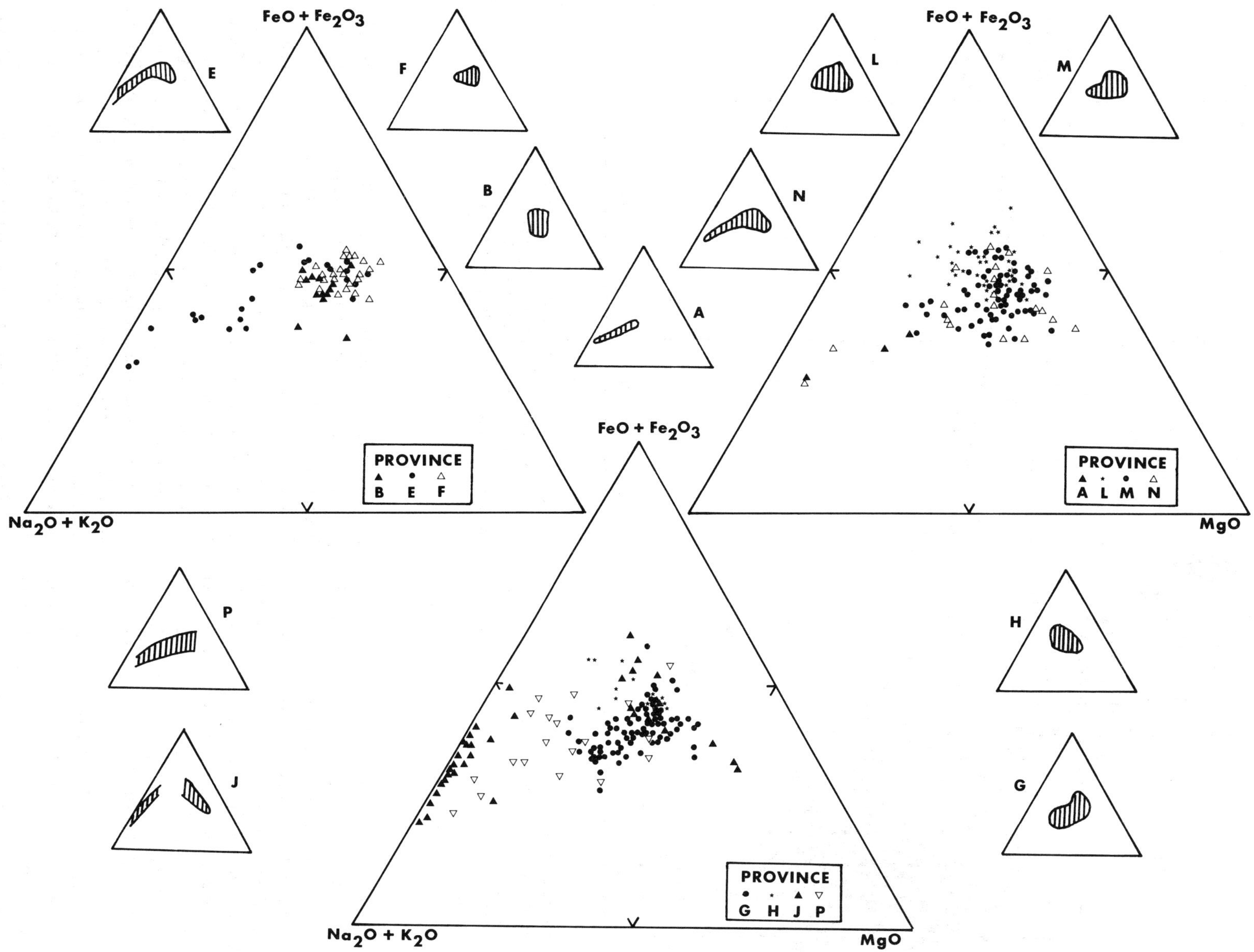

Figure 22. AFM diagrams for volcanic rocks of mainland Asia. Letters correspond to provinces outlined in Figure 20. Individual analyses are plotted in large triangles; small triangles indicate fields occupied by rocks from individual provinces. Most areas have very restricted compositional range. There is little evidence of any iron enrichment. Province J appears to exhibit classical "Daly gap." (Modified after Whitford-Stark [1983].)

1959b; Chasovitin, 1963; Gorodinskiy and others, 1967). The faults appear to have been originally of pre-Paleozoic age (Gorodinskiy and others, 1967) and have been frequently rejuvenated.

Province C. A similar association with rejuvenated faults appears to hold for the northern Siberia region, although the earliest volcanism may be associated with interactions between the American and Eurasian plates rather than to the Evenian belt (Andrews-Speed, 1981; Natal'in and Parfenov, 1983).

Province D. The Pleistocene-Holocene volcanics of Majak (listing 133) and Balagan-Tas (listing 132) are associated with a strike-slip fault zone caused by the relative rotations of the North American and Eurasian plates (Terman, 1977; Parfenov and Natal'in, 1979). Balagan-Tas is located near the crest of an anticline that is cut by north-northeast–trending faults (Argunov and Gavrikov, 1960).

Provinces E and F. The Udokan Plateau lies at the extreme northeast end of the Miocene–Early Pliocene Sayan-Baykal uplift; its location corresponds to a widening of the Chara rift depression—an extension of the Baykal rift (Kiselev and others, 1978). Geochemical data (Stupak, 1978a, b) suggest that either the younger volcanics were derived from greater depths than the older ones or that the earlier volcanics resulted from greater degrees of partial melting of the mantle.

The whole of the Sayan-Baykal uplift zone is underlain by a low-velocity, low-density asthenosphere with localized low-density anomalies under the north of the Udokan and Vitim plateaus and the East Sayan ridge (Kiselev and others, 1978). The lavas of the Vitim plateau lie off the uplift; this knowledge led Kiselev and others (1978) to suggest that volcanism and rifting in the region are independent processes and that instead of being directly related to rifting, volcanism is located in areas of anomalous mantle. Das and Filson (1975) argued that at the present time there is compression to the west of the southern end of Lake Baykal and extension in a zone from Baykal to the Sea of Okhotsk. This, they argued, resulted partly from rotation of the Chinese plate. Zonenshayn and Savotsin (1981) indicated an almost reverse situation with extension west of 56.3°N, 118.5°E and compression to the east. Tapponnier and Molnar (1977) attributed the formation of the Baykal rift to the India-Eurasia collision. Although the convergence of the two continents may have taken place over a period of 20 m.y. (Chandra, 1979), the main part of the collision appears to have taken place near the Eocene-Oligocene boundary (Molnar and Tapponnier, 1975). This would appear to correlate with the onset of volcanism in the Sayan-Baykal region and may have been the triggering event.

Province G. The volcanism of Mongolia has been controlled by faults (Devyatkin and Smelov, 1979). The most recent volcanics are associated with both east-west– and north-south–trending graben structures (Kepezhinskas and others, 1975). Some of the faults appear to follow former suture zones (Zonenshayn, 1967; Zonenshayn and Gorodnitskiy, 1977). The oldest basalts are located in the southernmost part of central Mongolia and become progressively younger to the north, with the youngest forming a latitudinal belt along approximately 48°N. Western Mongolia contains north-south–striking right-lateral strike-slip faults, whereas central Mongolia has left-lateral east-west–striking strike-slip faults, which Molnar and Tapponnier (1975) considered to be manifestations of the India-Eurasian collision.

The alkaline volcanics of Manchuria also appear to be related to deep-seated fault structures. This is particularly apparent at Wudalianchi (listing 20), where the volcanoes are arranged in a well-defined orthogonal pattern (see Fig. 5).

Provinces H, J, and L. During the Late Cretaceous, southeastern Korea was the site of calc-alkaline magmatism associated with a northward-dipping subduction zone (Sillitoe, 1977). The next event to affect the area was the creation of ocean floor between Japan and the mainland. Sillitoe (1977) placed this event as no earlier than mid-Eocene and cited other authors for evidence of a mid-Miocene separation. Lallemand and Jolivet (1986) have recently presented a model in which they envisaged the formation of the Sea of Japan by Eo-Oligocene rifting of a pull-apart basin, followed by the main opening phase during the Upper Oligocene/Lower Miocene as a mega pull-apart. This in turn was followed by an episode of back-arc spreading that induced the formation of the Yamoto Basin at the end of the Early Miocene. The basalts of the Yongil area (listing 1) probably correlate with this separation.

The Ch'uga-ryong (listing 3) and parallel Koksan-Singye (listing 2) belts are graben-related. Lee (1981) described the Ch'uga-ryong graben system that trends north-northeast–south-southwest as having been reactivated at least twice. The first stage, of Cretaceous age, was normal faulting with associated tholeiitic lavas. The lavas are overlain by a conglomerate, followed by a late Cretaceous tuff, which was associated with the second stage of faulting. The third phase of faulting was associated with the eruption of the Late Pleistocene–Holocene olivine basalts.

A similar scenario has been proposed for the subparallel North China Basin (Ye and others, 1985), where there was rifting during the Triassic through Jurassic, during the Cretaceous, and from the Eocene through Oligocene, and renewed activity during the Quaternary. Ye and others (1985) attributed the early Tertiary rifting to crustal extension and lithospheric thinning following uplift. This was followed by thermal subsidence. This scenario is similar to the basin-and-range type of tectonics found in North America and associated with crustal extension (e.g., Zhang and others, 1984).

Quaternary volcanism in this region also appears to be associated with, or at least parallels, large fault zones, particularly the Tancheng-Lujiang left-lateral fracture zone, that trends approximately north-northeast–south-southwest. Huzita (1984), however, reviewed information related to the Shangdong Province to the southeast of Bohai Bay and described Cretaceous sandstones thrust over Late Quaternary conglomerates as a result of horizontal compression. Alternatively, this could be an extensional detachment fault since there is an upward bulge in the mantle beneath Bohai Bay (Zheng Xuezheng, 1981; Zhang and others

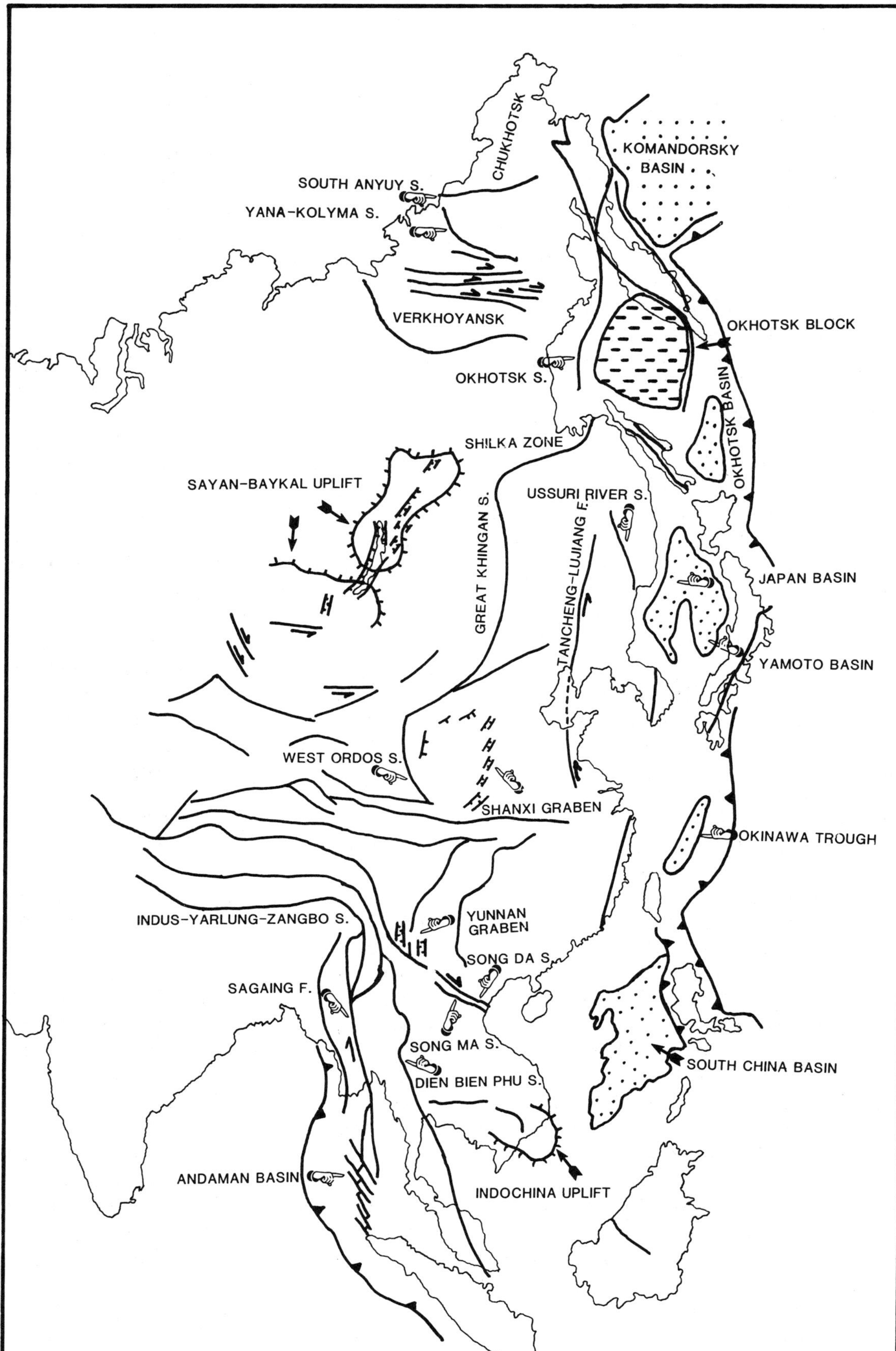
KOMANDORSKY BASIN
CHUKHOTSK
SOUTH ANYUY S.
YANA-KOLYMA S.
VERKHOYANSK
OKHOTSK BLOCK
OKHOTSK S.
OKHOTSK BASIN
SHILKA ZONE
SAYAN-BAYKAL UPLIFT
USSURI RIVER S.
GREAT KHINGAN S.
TANCHENG-LUJIANG F.
JAPAN BASIN
YAMOTO BASIN
WEST ORDOS S.
SHANXI GRABEN
OKINAWA TROUGH
INDUS-YARLUNG-ZANGBO S.
YUNNAN GRABEN
SONG DA S.
SAGAING F.
SONG MA S.
SOUTH CHINA BASIN
DIEN BIEN PHU S.
ANDAMAN BASIN
INDOCHINA UPLIFT

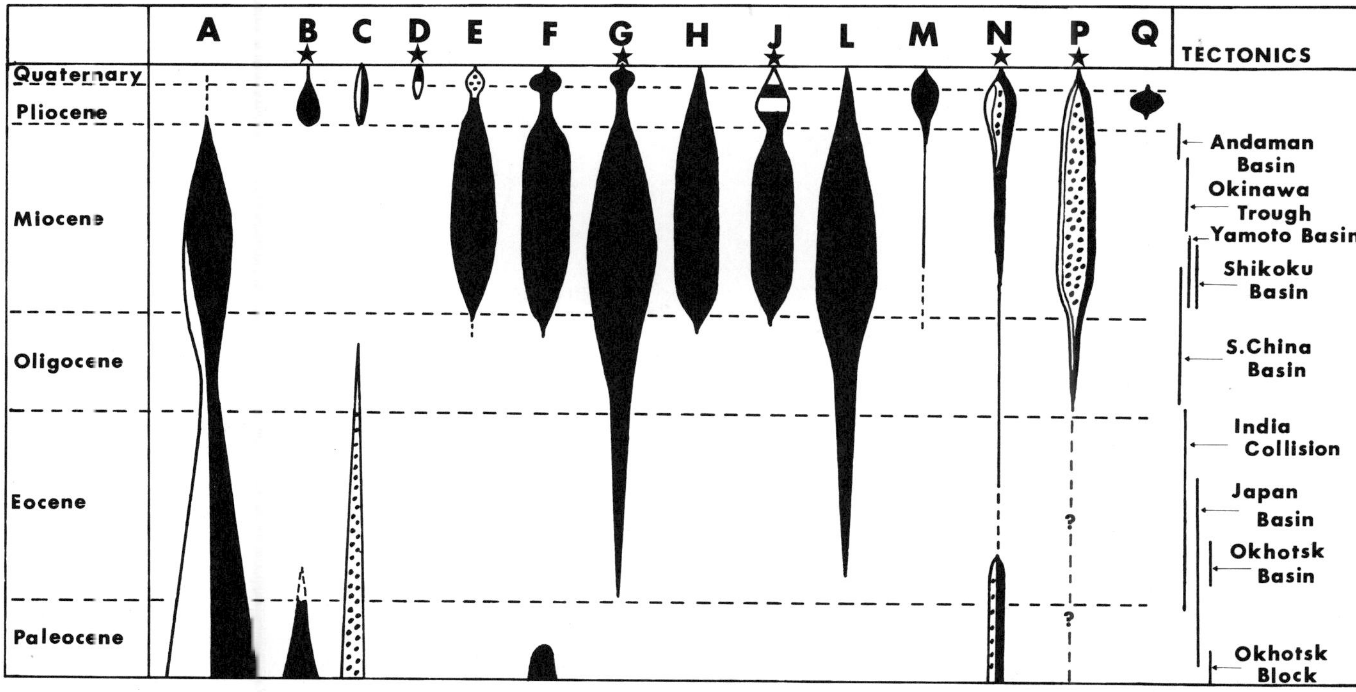

Figure 24. Illustration of volcanic stratigraphy and tectonic events of mainland Asia during Cenozoic. Letter designations are geochemical provinces outlined in Figure 19. Solid shading indicates basalts, dotted intermediate volcanics, and unshaded acid volcanics. Stars indicate historically active volcanism (inclusion of offshore islands would place province M in this category); in reality, volcanism more episodic than shown. Column widths reflect only variation in output within individual provinces, not variation in volume between provinces. Tectonic data from Chandra (1979), Crawford and others (1981), Curray and others (1979), Eguchi and others (1979), Molnar and Tapponnier (1975), Parfenov and Natal'in (1979), Taylor and Hays (1980), and Uyeda and Kanamori (1979). (Reprinted from Whitford-Stark [1983]; with permission of Elsevier.)

1984), which has led to a thinning of the crust and the development of a tensile fault system.

Province M. In southeast China there are extensive Mesozoic volcanic rocks (Wu Liren and others, 1983) and several pre-Cenozoic subduction zones. In contrast, Cenozoic volcanic activity has been minimal and along the extreme coastal zone, particularly on the Leizhou Peninsula (listing 46) and Hainan Island (listing 47). These relationships would suggest that the subduction zone had moved seaward from the continent by the end of the Cretaceous. The migration of the subduction zone resulted from the formation of the South China Basin dated as beginning in the Late Jurassic to Early Cretaceous by Ben-Avraham and Uyeda (1973) and Late Oligocene to Early Miocene by Jurdy (1979). Likewise, farther north, the formation of the Okinawa trough sometime after Late Miocene-Early Pliocene (Herman and others, 1979) probably exerted a similar influence.

The isolated Tengchong area (listing 49) of Yunnan may be related to a subducted Indian Ocean plate, although the earthquake foci of the Burma arc do not currently extend to the Tengchong area (Zhu and others, 1983). Zhu and others (1983) proposed that the late Tertiary volcanics were erupted in a tensional environment and that this changed to a compressional environment during the Early Pleistocene and back to an extensional environment in the Late Pleistocene.

In Thailand there appears to have been a Permo-Triassic arc associated with eastward subduction during the Permian; the suture zone representing the closure of the ocean or marginal basin is Triassic in age (Beckinsale and others, 1979). Likewise the collision of the Thai-Malay and Indochina blocks with the South China block is believed to have taken place during the Middle or Late Triassic (Ridd, 1980).

Barr and Macdonald (1978) suggested that the basalts of southeast Asia may be related to the opening of the South China

Figure 23. Map of some major tectonic features of mainland Asia. Emphasis on features described in text that have been active in influencing distribution of Cenozoic volcanism. Different authors have placed their boundaries at different locations; thus map biased toward more recent publications. S = suture; F = fault. Map based on Beckinsale and others (1979), Burrett (1974), Chingchang (1979), Curray and others (1979), Das and Filson (1975), Fujita and Newberry (1983), Hsieh (1963), Karig (1971), Klimetz (1983), Krasnyy (1967), Matsuda and Uyeda (1971), Mitchell and McKerrow (1975), Parfenov and Natal'in (1979), Parfenov and others (1981), Ridd (1980), Rodolfo (1969), Sengor (1984), Tapponnier and Molnar (1977), Zhang and others (1984).

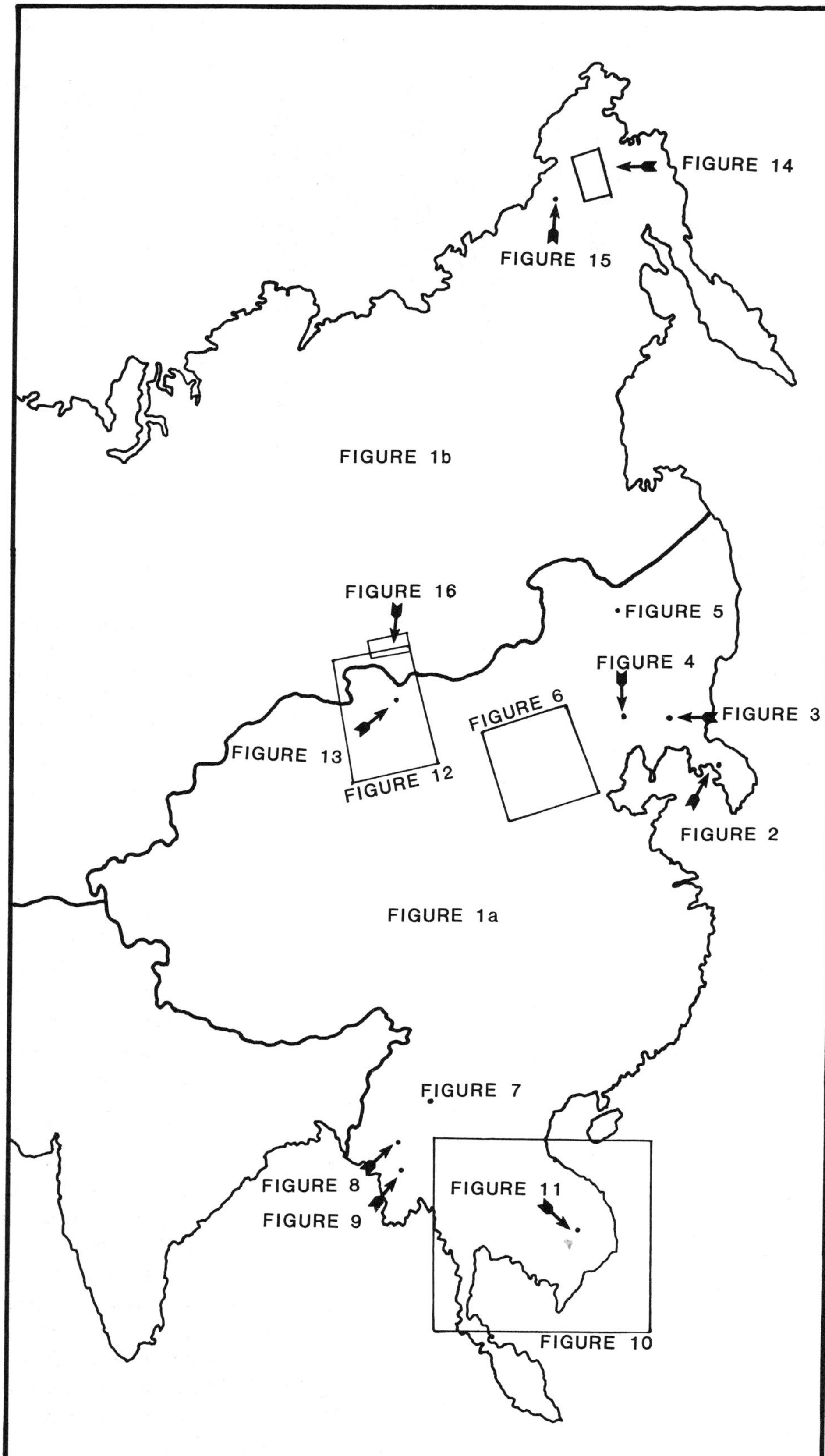

Figure 25. Index map showing location of figures in text.

Basin, but the relatively young age of the basalts (largely Plio-Pleistocene) and the early termination of formation of the South China Basin during the Miocene (Ben-Avraham and Uyeda, 1973) do not support this view. The only event of approximately the same age as the basalts is the opening of the Andaman Basin. The problems associated with relating to southeast Asian volcanics to the Andaman Sea spreading include those of distance (1,300 km), an increase in volume and decrease in age to the southeast; furthermore, the trends of the fissures (Saurin, 1967; Lacombe, 1969) do not coincide with trends of Andaman Sea features. A possible alternate argument is that the basalts are related to a local "hot spot" (Whitford-Stark, 1983). Possible support for such an argument is found in doming of the area (Lacombe, 1969), in high heat-flow values recorded in the western part of the South China Basin (Watanabe and others, 1977), and in the occurrence of historic volcanic activity in the offshore islands (Simkin and others, 1981). Yet, there is no simple age-location progression; old and young volcanics are found in adjacent areas. Tapponnier and others (1982) indicated that the clockwise rotation of Indochina as a result of collision of India with Eurasia would have been most manifest during the period from 50 to 20 m.y. B.P.; thus this rotation does not appear to correlate directly with the age of the magmatic activity.

Province N. The geochemistry of the Burmese volcanics is consistent with their derivation as high-K calc-alkaline continental margin magmas (Stephenson and Marshall, 1984). These rocks were largely absent until the Miocene, and this has led Mitchell and McKerrow (1975) to suggest that this could result from nonsubduction, slow subduction, or tectonic emplacement of ocean floor rocks in the Indo-Burma ranges, causing subduction of only small amounts of oceanic crust. Fault plane solutions indicate that underthrusting is not taking place at the present time, but intermediate depth seismic events suggest that a hanging slab is being dragged north by India through the surrounding asthenosphere (Le Dain and others, 1984).

The apparent increase of volcanism during the Pliocene may be related to the northward movement of the West Burma block (Mitchell, 1981) or to the opening of the Andaman Basin (Rodolfo, 1969; Curray and others, 1979; Eguchi and others, 1979). The northward motion of India is currently being accommodated by dextral movement on the Sagaing fault system, which is estimated to have amounted to as much as 450 km since the early Miocene (Mitchell, 1981). To the east of the Sagaing fault, deformation is currently dominated by conjugate strike-slip faulting in response to north-south shortening and east-west extension; this results from the eastward expulsion of Asia out of the path of India (Le Dain and others, 1984).

Province P. Deng Wanming (1978a) has attributed the chemical variation of the northern Tibetan block volcanics to differnt depths of origin of materials derived from the southward underthrusted Tarim block. An alternate model, suggested by Jin Chengwei (1981), is that the volcanism is related to movement along the Kunlun fault. To the south, the Tibetan block has been thrust over the Indian continent (e.g., Zhang and others, 1984), with resultant magmas being derived from both within the crust and the mantle (Sengor and Kidd, 1979). Zhou (1985) presented evidence for five distinctive, subparallel magmatic zones, trending west-northwest–east-southeast, within Tibet: a central zone related to convergence of the Indian continent, two northern zones related to earlier collisions of the Eurasian and other microplates, and two southern zones postdating the Indian continental collision. The magmatic rocks in the central zone apparently are progressively younger to the north and to the east. Zhou (1985) ascribed the eastward younging to a diachronous collision.

Province Q. These isolated mainland occurrences of ultrapotassic lavas could be subduction-related, maintaining the trend (Ninkovich and Hays, 1972) of increasing alkalinity with distance from the Indonesian arc.

VOLCANOTECTONIC SUMMARY

Although the India-Eurasia collision has exerted a profound influence over the timing and location of volcanism throughout mainland Asia, the volcanism cannot be entirely ascribed to the effects of that event. It is also necessary to take into account the reactivation of earlier plate boundaries, the opening of marginal back-arc seas and crustal thinning, and interactions with the North American plate. Changes in plate motion have been reflected on mainland Asia by the initiation or termination of volcanism or by a change in style or composition of volcanism.

The Eurasian plate has undergone an evolution somewhat similar to that of the western North American plate, which has involved construction from the suturing of a large number of small plates. These plate and microplate boundaries have frequently acted as zones of weakness, permitting the eruption of volcanic materials far into the continental interior.

APPENDICES

Note: A geographic glossary is provided to assist readers who are unfamiliar with the various Asian languages. A rock glossary includes some commonly used rock names but also includes names that have passed from common usage and others which are confined to the Asian literature. Tables of rock types and volcano types are provided for quick reference to the catalog. The volcano dimensional data were used in compiling Figure 17 of the text.

The many geochemical analyses that are listed often were not made by the author indicated in the reference and some are fairly old. Additionally, some are averages of a large number of samples. The reader is referred to the quoted author for details of individual analyses. Values for "calculated loss" and "volatile loss" have not been included. The decimal point preceding the Nd and Sr isotope ratios has been omitted, as have the names originally assigned to the individual specimens.

APPENDIX 1. GEOGRAPHIC GLOSSARY

Ban (Laos) = Village
Bei (Ch.) = North
B'nom (Viet.) = Mountain
Bo (Ch.) = Lake
Bogd (Mon.) = Range

Cao nguyen (Viet.) = Plateau
Chen Tou (High.) = Mountain
Chi (Ch.) = River
Chi (Ch.) = Lake
Ch'on (Kor.) = River
Chrebet (Sov.) = Range
Chu (High.) = Mountain
Ch'uan (Ch.) = River
Co (Viet.) = Mountain

Da (High.) = Stream
Da (Ch.) = Large
Dao (Ch., Viet.) = Island
Do (Kor.) = Island
Doi (Th.) = Mountain
Dong (Ch.) = East

Feng (Ch.) = Mountain

Gang (Kor.) = River
Gol (Mon.) = River
Gon (Ch.) = River
Gory (Sov.) = Mountains
Gora (Sov.) = Mountain

He (Ch.) = River
Ho (Ch.) = River
Hsi (Ch.) = West
Hsiao (Ch.) = Small
Hu (Ch.) = Lake
Huo (Ch.) = Fire
Huoshan (Ch.) = Volcano
Huoshanyan (Ch.) = Volcanic rock

I (Ch.) = Armchair

Jary (Sov.) = Mountains
Jiang (Ch.) = River

Kang (Kor.) = River
Khao (Bur., Th.) = Mountain
Kou (Ch.) = Mouth

Long (Ch.) = Dragon
Lung (Ch.) = Dragon

Nagorje (Sov.) = Plateau
Nam (High.) = Stream
Nan (Ch.) = South
Ngoc (Viet.) = Mountain
Noy (Laos) = Small
Nui (Viet.) = Mountain
Nuruu (Mon.) = Mountains
Nuur (Mon.) = Lake

O (Kmer) = Stream
Ozero (Sov.) = Lake

Palei (High.) = Village
Pei (Ch.) = North
Phnom (Kmer) = Mountain
Phou (Laos) = Mountain
Phu (Th.) = Mountain
Phum (Kmer) = Village
Plato (Sov.) = Plateau
Plei (High.) = Village
Ploskogorje (Sov.) = Plateau
Pong (Kor.) = Peak
Prok (Kmer) = River

Rach (Viet.) = River
Rongyan (Ch.) = Lava

San (Kor.) = Mountain
Sanmaek (Kor.) = Mountains
Se (Laos) = River
Shan (Ch.) = Mountain
Shi (Ch.) = Stone
Shui (Ch.) = River
Song (High.) = Stream
Sopka (Sov.) = Volcanic cone
Stung (Kmer) = River
Su (Kor.) = River

Ta (Ch.) = Large
Tang (Ch.) = Barrier
Touch (Laos) = Small
Tung (Ch.) = East

Ula (Mon.) = Mountain
Uul (Mon.) = Mountains

Wan (Ch.) = Pit

Xi (Ch.) = River
Xi (Ch.) = West
Xiao (Ch.) = Small
Xuanwuyan (Ch.) = Basalt

Ya (High.) = River

Key to abbreviations: Bur. = Burmese, Ch. = Chinese, High. = Montagnard or highland, Kor. = Korean, Laos = Laotian, Mon. = Mongolian, Sov. = Soviet, Viet. = Vietnamese.

APPENDIX 2. ROCK GLOSSARY

Andesite: An intermediate extrusive rock with andesine or labradorite and some combination of augite, orthopyroxene, or hornblende. Some may have bytownite; others, olivine.

Ankaramite: An olivine-bearing basalt with pyroxene and olivine phenocrysts. A fine-grained matrix of clinopyroxene micolites and calcic plagioclase.

Basanite: An undersaturated trachybasalt with olivine, calcic plagioclase, clinopyroxene, and a feldspathoid (usually nepheline or leucite). Without olivine it would be a tephrite.

Comendite: A peralkaline rhyolite or quartz trachyte, less mafic than a pantellerite.

Hakutoite: A rock named by Ogura (1969) from Hakuto (P'aektu-san). A fayalite-soda pyroxene-aegirine augite-anorthoclase pantellerite. Grayish-blue with abundant phenocrysts of anorthoclase and augite in a compact groundmass.

Hawaiite: Olivine "basalt" with normative andesine. Soda: potash ratio greater than 2:1. Generally lacks normative quartz and commonly has normative and modal olivine.

Kokuhangan: A black porphyritic basalt rich in potassium (Tomita, 1969).

Kulaite: A nepheline trachybasalt or nepheline basanite. Hornblende is the dominant mafic mineral (Washington, 1900). Not a recommended term.

Latite: A porphyritic rock with phenocrysts of plagioclase and K-feldspar in nearly equal amounts and little or no quartz. A finely crystalline to glassy groundmass. Synonymous with trachyandesite and trachybasalt.

Limburgite: Olivine and augite in an ultrabasic, alkaline, glassy groundmass. No feldspar. Not a recommended term.

Liparite: Synonymous with rhyolite. Used to indicate a subaluminous

rhyolite. Not a recommended term.
Mugearite: An alkali basalt. A trachybasalt or trachyandesite with oligoclase as the feldspar.
Pantellerite: A peralkaline rhyolite with normative quartz exceeding 10 percent. More mafic than a comendite.
Perknite: Any group of igneous rocks with amphibole and clinopyroxene as its major constituents with accessory orthopyroxene, biotite, iron oxide, and little or no feldspar. Not in common usage.
Saxonite: A peridotite composed chiefly of olivine and orthopyroxene. Approximately synonymous with harzburgite.
Shihlunite: Also spelled shiklunite. A leucite basanite. Black obsidian-like rock with 80 percent glass and microphenocrysts of leucite, olivine, augite, and anorthoclase (Ogura, 1969).
Tephrite: An undersaturated trachybasalt without olivine. Nepheline and leucite are the main feldspathoids.
Teschenite: Granular rock with calcic plagioclase, augite, sometimes hornblende and a small amount of biotite, with interstitial analcime.
Trachyte: Generally porphyritic with alkali feldspar and minor mafics (biotite, hornblende, or pyroxene) as the main constituents. Also smaller amounts of sodic plagioclase. Trachyandesite and trachybasalt are intermediate in composition between basalt, andesite, and trachyte. A trachyandesite is approximately synonymous with a mugearite and trachybasalt with hawaiite.
Wehrlite: A peridotite composed chiefly of olivine and clinopyroxene with common accessory opaque oxides.

APPENDIX 3. LOCATIONS OF ROCK TYPES BY CATALOG NUMBER

Analcitite: 50
Andesite: 1, 47, 49, 50, 55, 57, 59, 60, 63, 66, 124, 134, 139, 154, 155
Ankaramite: 72
Augitite: 122, 124

Basalt: 2, 4, 5, 9, 12, 17, 18, 21, 22, 23, 24, 30, 31, 32, 33, 34, 37, 38, 39, 40, 42, 45, 46, 47, 54, 56, 57, 60, 62, 68, 76, 87, 88, 89, 90, 92, 93, 94, 95, 96, 97, 98, 99, 100, 101, 102, 103, 104, 105, 106, 107, 108, 109, 110, 111, 112, 113, 114, 115, 116, 117, 118, 119, 120, 121, 126, 127, 132, 136, 140, 141, 143, 144, 148, 149, 150, 151, 152, 159, 161, 164, 167, 168, 169, 170
- **analcite-:** 4, 123, 124, 125, 153, 154
- **andesite-:** 122, 125, 129, 130, 134
- **biotite-:** 123
- **carbonatite-:** 160
- **leucite-:** 20, 124, 142
- **nepheline-:** 11, 41, 77, 84, 145, 146, 147
- **oligoclase-:** 125
- **olivine-:** 1, 3, 4, 5, 6, 7, 8, 10, 13, 14, 15, 16, 20, 25, 35, 36, 38, 43, 44, 49, 56, 58, 63, 64, 65, 66, 70, 71, 76, 77, 78, 83, 85, 86, 116, 156, 159, 160, 162, 163, 165, 166
- **picrite-:** 40, 45, 58, 64, 125
- **sanidine-:** 123
- **tholeiitic-:** 5, 7, 46, 47, 56, 73, 88, 108
- **trachy-:** 4, 7, 20, 48, 86, 122, 123, 124, 125, 127, 128, 131, 154, 155

Basanite: 7, 10, 13, 14, 20, 71, 72, 74, 84, 122

Clinopyroxenite: 14
Comendite: 4

Dacite: 50, 55, 63, 134, 139, 141
Diorite: 63
Dolerite: 56, 57, 63, 67, 68, 150
Dunite: 13, 125

Hakutoite: 5
Harzburgite: 7, 125
Hawaiite: 7, 75, 76, 77, 79, 80, 81, 83, 122

Ignimbrite: 55, 66, 126, 136, 152

Kokuhangan: 40
Kulaite: 153

Latite: 13, 54, 66
Lherzolite: 7, 14
- **spinel-:** 13, 76, 77, 122, 125

Limburgite: 28, 91, 122, 123, 124, 125, 127
Liparite: 26, 27, 133, 134, 137, 138, 139

Monzonite: 54
Mugearite: 61, 80, 82, 83

Nordmarkite: 5

Peridotite: 58, 63, 122, 153
Perknite: 63
Phonolite: 13, 50
Porphyry: 63
Pseudoleucite: 50
Pyroxenite: 7, 63, 125

Rhyodacite: 66
Rhyolite: 50, 55, 58, 63, 69

Saxonite: 13
Serpentinite: 58
Shihlunite: 20, 158, 169

Tephrite: 13, 72, 122, 127, 129
Teschenite: 61
Trachyte: 4, 5, 27, 60, 134

Wehrlite: 125

APPENDIX 4. AGES OF VOLCANISM, LOCATED BY CATALOG NUMBER

Historic: 5, 20, 28, 50, 51, 53, 54, 66, 127, 128, 132
Pleistocene–Holocene: 3, 5, 6, 7, 8, 9, 10, 12, 13, 17, 18, 19, 20, 22, 23, 24, 25, 26, 28, 34, 37, 45, 46, 49, 61, 63, 74, 84, 86, 96, 97, 111, 114, 115, 116, 122, 125, 132, 133, 135, 136, 152, 153, 154, 155, 156, 158, 159, 160, 161, 162, 163, 165, 166
Pleistocene: 2, 3, 4, 5, 13, 20, 27, 32, 36, 43, 47, 48, 49, 66, 73, 81, 85, 87, 88, 89, 90, 91, 92, 93, 94, 95, 98, 99, 100, 101, 102, 103, 104, 105, 106, 107, 108, 110, 113, 114, 116, 117, 118, 121, 122, 129, 130, 131, 134, 165, 168
Pleiocene–Pleistocene: 42, 49, 55, 58, 63, 65, 69, 70, 109, 112, 117, 119, 120, 125, 152
Pliocene: 4, 5, 13, 49, 62, 66, 71, 72, 77, 83, 84, 91, 99, 101, 107, 114, 122, 137, 138, 139, 140, 141, 142, 143, 144, 145, 146, 147, 148, 149, 150, 151
Miocene–Quaternary: 14, 15, 16, 39, 56, 57, 60, 67, 68, 153, 157, 158, 169
Miocene–Pliocene: 40, 55, 76, 84, 159, 160
Miocene: 1, 4, 30, 63, 78, 103, 116, 123, 152
Oligocene–Miocene: 60
Oligocene: 35, 38, 60, 84, 123, 152, 159, 165, 167
Eocene–Miocene: 45, 55
Eocene: 40, 124, 152
Paleogene: 72, 126, 152
Unknown: 11, 21, 31, 41, 44, 59, 64, 75, 79, 80, 82, 164, 170

APPENDIX 5. VOLCANO TYPES, LOCATED BY CATALOG NUMBER

Calderas: 5, 55, 96, 99, 103, 104, 109, 126, 141
Composite: 5, 50, 51?, 53?, 54?, 57, 59?, 66, 155
Cones (cinder): 5, 6, 7, 18, 19, 20, 22, 23, 25, 27, 28, 29, 31, 32, 33, 34, 36, 42, 43, 46, 47, 49, 50, 84, 86, 93, 94, 95, 96, 98, 99, 107, 109, 110, 111, 113, 114, 115, 117, 118, 119, 120, 121, 122, 125, 127, 128, 129, 132, 140, 142, 143, 144, 145, 150, 151, 153, 157, 158, 160, 161, 163, 164, 165, 168, 169
Cones (lava): 13, 15, 21, 99, 105, 116, 125
Domes: 5, 9, 10, 17, 60, 65, 66, 133, 137, 138, 139, 141, 146, 147
Explosion: 26, 48, 63, 64, 86, 114, 135, 162
Fissures: 26, 58, 61, 71, 86, 128, 156, 166
Intrusions: 67, 68, 85, 122
Necks: 63, 64, 65, 77, 83, 122, 123, 124
Plateaus: 5, 12, 14, 16, 19, 20, 28, 30, 35, 43, 55, 56, 58, 61, 63, 70, 71, 86, 87, 88, 89, 90, 92, 99, 100, 101, 103, 107, 108, 117, 123, 125, 136, 157, 169
Shields: 2, 3, 5, 8, 15, 20, 48, 54, 56, 102, 126, 141, 148, 149, 154, 165, 168
Solfataras: 52
Table Mounts: 50?
Unknown: 11, 37, 39, 41, 44, 45, 69, 73, 75, 112, 130, 131, 134, 135, 152, 170

Note: Many of the features listed as calderas in the Post-Miocene Volcanoes of the World catalog (IAVCEI, 1973) are much smaller than the 1-km diameter lower limit usually recognized for such features (cf. Basaltic Volcanism Study Project, 1981, p. 745) and are probably maars. Likewise, features described as "destructive craters" could be cinder cones or maars, as are explosion craters.

APPENDIX 6. VOLCANO DIMENSIONS

Name	Locality	Height (m)	Width (m or km)	Crater Diameter (m or km)	Crater Depth (m)	Type
Ap-san	3	100		200	20	shield
Dayenzhifeng*	5	30	1000			cone
Paektu-san	5**	1750	40 km	5 km	869	cone
Xiaoyenjifeng*	5	30	1000			cone
Dayishan (ta-i-)	6	179	524			cone
Xiaoyishan (Hsiao-i-)	6		455			cone
Huangyishan (Huang-yi)	7	100				cone
Datugeerjishan	10	44	700			cone
Donghalabanshan	10	90	555			dome
Naobaoshan (Noa-pao-)	10	30	320			dome
Polishan	10	110	770			dome
Popotushan	10	70	900			dome
Xiaotugeerjishan	10	50	490			dome
Xihalabashan	10	70	650			dome
Dongshan	19	110	700	230	35	cone
Xiaokeshan	19	18	210	80	5	cone
Xishan	19	75	550	250	50	cone
Beigelaqiushan	20	57.5	550	230		cone
Bijiashan (Pi-chia)	20	125.3	1000	400	70	cone
Dongjiaodebushan	20	106.9	750	340	43	cone
Donglongmenshan	20	137.4	950	430	80	cone
Huoshaoshan (Hu-shao-)	20**	70.3	750	370	70	cone
Laoheishan (Lao-hei)	20**	165.9	1300	350	145	cone
Molabushan	20	122.8	1500	250	40	cone
Nangelaqiushan	20	146.9	1000	500	50	cone
Weishan	20	103	800	320	41	cone
Wohushan	20	116.1	1200	400	35	cone
(double cone)		78.8		300	20	cone
Xiaogushan	20	73.5	620	330	35	cone
Xijiaodebushan	20	99.8	650	250	25	cone
Xilongmenshan	20	156.4	1000	350	80	cone
Yaoquanshan (Yao-chuan)	20	60.8	500	240	30	cone
Daheishan	22				100	cone
Dianzishan	22	200				cone
Keluonanshan	22	120				cone
Xiaoyishan	22	50				cone
Xishan	22	230				cone
Shadeershan	25	100		400		cone
Cone 1	25	300				cone
Cone 2	25	250				cone
Cone 8	25	100		200	50	cone
Ghenor	26	250-300	2000			explosion
Baerqian	27	200	1000			composite?
Nyushan (Nu-shan)	42	55				cone
Weizhoudao	48		4-5 km	1000		shield
Dayingshan (Tay-in)	49**	990	7.5 km			composite?
Laoguibo	49	244		182	30	cone
Maanshan	49	275	9.6 x 6.4 km			cone
Unnamed	50			4000		caldera
Unnamed	50		1500			cone
Bamaoqionzong***	50	150		230	20+	cone
Qiangbaqian***	50	200				table?
Mt. Loimye	57	1000				composite
Ywatha	63			850	45	explosion
Taungbyauk	63			800	60	explosion
Twindaung	63	150		1450	143	explosion
Twinywa	63			1200	61	explosion
Mt. Popa	66**	1150	7.2 km	1.6 km	850	composite
Lom Kot	86			600 x 300		explosion?
Ya Lom	86			800	60	explosion?
Ya Om	86			200		explosion?
Ya Ra	86			200		explosion?
Cu-lao Bo Bai	96	20				cone
Troeng Prong	98	60				cone?
Chi-Haedrong	99	122				"caldera"
Chi-Holam	99	60				lava cone
Chi-Ten	99	70				"caldera"
La Bang	99	55				"caldera"
Pleiku	99	60				"caldera"
Plei-Monu	99	80				"caldera"
Pleitott	99	40				lava cone

*Ogura (1969) assigned these values to unnamed cinder cones.
**Indicates volcano reported as historically active.
***Name of area other than volcano.

Note: Volcanoes labeled explosion and "caldera" are probably maars.

APPENDIX 6. (continued)

Name	Locality	Height (m)	Width (m or km)	Crater Diameter (m or km)	Crater Depth (m)	Type
Lien Khang	104	10				"caldera"
Suzanna	105	180				lava cone
Cam Tiem	106	100				lava cone
Nui Dat	107	77				cone
Khutza-Ula	109	162				cone?
Nomogan-Ula	109	21				cone?
Tabun-urta-ula	110	67				cone?
Bus Obo	114	90				cone
Bogo-togo-ula	115	28				cone?
Ika-togo-ula	115	219				cone?
Durbulzhi-ula	120	45				cone?
Khorog	122	120	1200	180		cone
Unnamed	122	160	1000	400		cone
Gremuchaya	126			25 km		caldera
Kyrgannivaam	126		60 km			shield
Marakveem	126		50 km			sheild
Pyrkhanaiveem	126			20 km		caldera
Shchitovaya	126	300	25 x 75 km			shield
Terpukhoiskii	126		55 km			shield
Varynnaiskaya	126			20 km		caldera
Alpha	127	70	450	130	30	cone
Beta	127	40	70?	20		cone
Gamma	127**	100	400			cone
Aniusk	128**	120	490?	350	75	cone
Bilibina	129	390		160		cone
Balagan Tas	132**	300	1200	200	40	cone
Shishlovsky	140	40				cone
Uysky	141	300				shield
Bolonsky	142	15				cone
Sindinsky	143	32				cone
Sviatogorsky	144	68				cone
Imansky	145	56				cone
Medvegiy	146	247				dome
Ostry	147	100				dome
Sviaghinsky	148	30				shield
Merkushovsky	149	156				shield
Baevsky	150	113				cone
Razdolninsky	151	258				cone
Dvoinoi	153	15	250 x 600			dome
West Sakukan	153	30	250 x 600			cone
Yakutsk	153	30	550 x 700			dome
Chepe	154	154				shield
Aku	155	220				somma
Sini	156	250				cone?
Khara-Boldock	160	125				cone
Svyashchennaya Gora	160	40				cone
Kropotkin	163	90				cone
Peretolchin	163	120				cone
Strariy	163	40				cone
Derbi-Tayga	168	700				shield

*Ogura (1969) assigned these values to unnamed cinder cones.
**Indicates volcano reported as historically active.
***Name of area other than volcano.

Note: Volcanoes labeled explosion and "caldera" are probably maars.

APPENDIX 7. GEOCHEMICAL ANALYSES

	Locality 3 Ch'uga-ryong												Locality 4 Kilchu-Myongch'on		
SiO_2	49.84	47.38	46.59	46.44	48.44	47.23	47.10	46.80	47.00	47.00	46.77	47.22	50.89	43.48	43.52
Al_2O_3	16.33	15.11	16.07	15.50	11.89	14.73	14.39	15.36	15.83	14.57	13.93	15.71	17.28	16.78	9.83
Fe_2O_3	2.32	1.20	2.60	1.78	3.60	3.22	4.83	2.07	0.81	0.61	0.70	1.41	1.75	2.71	2.48
FeO	7.63	9.18	7.94	8.89	7.74	7.21	5.80	8.56	9.29	9.40	10.19	8.40	7.34	6.20	6.96
MgO	4.80	11.12	10.64	11.27	11.90	11.32	11.50	11.35	10.58	10.80	12.24	10.58	4.88	10.66	14.09
CaO	9.82	8.48	8.80	8.44	8.96	9.04	8.32	8.36	8.50	8.48	8.45	7.87	11.56	10.24	13.36
Na_2O	1.44	3.23	3.06	3.11	3.53	3.08	3.18	3.43	3.20	3.15	3.37	3.08	2.77	3.21	3.22
K_2O	3.51	2.00	1.56	1.94	1.80	1.62	1.97	1.68	1.37	1.39	1.89	1.55	0.71	0.69	0.87
H_2O+		0.33	0.35	0.48	----	0.20	0.30	----	0.72	0.32	0.19	0.90	0.10	2.79	2.23
H_2O-	0.18	0.36	0.37	0.36	0.19	0.34	0.57	0.27	0.56	0.65	0.19	1.07	0.29	0.54	0.31
TiO_2	3.95	1.52	1.59	1.40	1.67	1.62	1.47	1.55	2.00	2.12	1.39	2.10	1.95	1.04	1.01
P_2O_5		0.40	0.26	0.34	0.26	0.32	0.37	0.27	0.26	0.25	0.36	0.29	0.23	1.05	2.63
MnO	0.02	0.16	0.17	0.15	0.18	0.18	0.16	0.19	0.18	0.19	0.16	0.17	0.81	0.17	0.17
BaO															
CO_2															
Ba															
Sr															
Rb															
Zr															
Y															
La															
Ce															
Nd															
Sm															
Nb															
Li															
B															
U															
Th															
Mo															
Pb															
Zn															
Cu															
Ni															
Cr															
V															
Co															
$^{87}Sr/^{86}Sr$															
$^{143}Ne/^{144}Nd$															
AGE	[illegible]	Q	Q	Q	Q	Q	Q	Q	Q	Q	Q	Q			
Reference	1	2	2	2	2	2	2	2	2	2	2	2	1	1	1

APPENDIX 7 (continued)

	Locality 4 Kilchu-myongch'on														
SiO_2	44.32	44.90	47.82	49.52	63.52	67.88	68.04	68.17	72.54	73.13	73.21	73.38	74.28	74.30	74.65
Al_2O_3	12.51	18.57	18.85	16.42	15.80	13.47	12.83	12.47	11.79	12.78	14.55	11.88	12.48	10.40	10.43
Fe_2O_3	3.68	7.89	4.96	3.53	3.47	2.46	2.50	0.05	2.82	2.87	0.40	2.32	0.74	2.75	3.38
FeO	5.27	4.20	6.06	5.56	2.89	3.27	3.72	0.98	1.91	1.29	2.39	1.91	2.03	2.15	1.30
MgO	13.98	5.11	4.85	6.52	0.45	tr.	0.80	0.50	tr.	0.22	1.22	tr.	0.17	0.37	0.08
CaO	10.82	8.22	8.52	7.76	1.23	1.14	3.23	0.66	0.52	0.05	1.10	0.54	0.70	0.32	0.46
Na_2O	3.15	3.02	2.52	2.94	5.65	5.72	3.99	7.89	4.44	3.71	2.69	4.08	2.75	4.19	4.13
K_2O	0.80	3.33	2.43	2.35	5.50	4.35	5.48	6.10	4.87	5.14	3.80	5.05	5.81	5.03	4.97
H_2O+	2.71	1.62	1.60	1.74	0.43	0.57	0.58	2.82	0.53	0.25	0.38	0.30	0.91	0.50	0.21
H_2O-	1.04	0.52	0.95	1.09		0.67	0.32	1.10	0.27	0.16	0.18	0.14	0.17	0.40	0.25
TiO_2	1.18	1.03	1.47	1.32	tr.	0.36		tr.	0.40	0.23	0.06	0.28	0.43	0.37	0.23
P_2O_5	0.57	0.49	tr.	1.09	tr.	0.07		tr.	0.06		0.01	0.08		0.04	0.01
MnO	0.20	0.79	0.14	0.15	0.05	0.18	tr.		0.13	0.05	0.23	0.10	tr.	0.09	
BaO															
CO_2															
Ba															
Sr															
Rb															
Zr															
Y															
La															
Ce															
Nd															
Sm															
Nb															
Li															
B															
U															
Th															
Mo															
Pb															
Zn															
Cu															
Ni															
Cr															
V															
Co															
$^{87}Sr/^{86}Sr$															
$^{143}Ne/^{144}Nd$															
AGE															
Reference	1	1	1	1	1	1	1	1	1	1	1	1	1	1	1

APPENDIX 7 (continued)

	Locality 4 Kilchu-Myongch'on						Locality 5 P'aektu-san								
SiO_2	75.18	68.17	69.88	70.19	73.26	74.27	47.52	65.84	67.42	67.68	70.81	77.76	72.76	73.20	50.43
Al_2O_3	11.40	16.77	15.50	15.17	15.03	16.92	15.67	15.39	15.07	12.55	10.63	11.65	11.06	9.31	15.18
Fe_2O_3	1.25	2.44	3.66	2.01	0.96	0.98	7.81	1.45	0.44	3.00	2.60	2.51	2.94	3.25	2.77
FeO	2.45	2.00	2.12	2.46	1.46	2.28	4.47	3.55	2.79	2.85	4.01	3.23	2.46	3.10	8.92
MgO	tr.	0.12	0.36	0.28	tr.	0.51	3.78	0.03	0.25	0.24	0.23		0.10	0.04	4.98
CaO	0.48	0.42	0.27	0.24	tr.	1.30	8.18	1.50	1.42	1.25	0.75	0.62	0.59	0.38	8.96
Na_2O	3.92	2.99	1.96	1.76	3.34	3.26	4.17	5.96	6.11	6.31	5.97	4.68	4.55	5.25	3.13
K_2O	4.68	4.55	3.74	5.74	5.06	0.51	2.22	5.27	5.01	4.71	3.47	4.51	4.33	4.43	1.52
H_2O+	0.30	0.87	0.91	0.38		0.41	0.68	0.21	0.36	0.51	0.93	0.59	0.71	0.28	0.01
H_2O-	0.35	0.40	0.55	0.26	0.46	0.19	0.30	0.08	0.05	0.08	0.40	0.22	0.33	0.12	0.11
TiO_2	0.24	0.25	0.16				4.16	0.66	0.72	0.31	0.11	0.38	0.29	0.60	2.90
P_2O_5	0.06	tr.	tr.	tr.	tr.	tr.	0.76	0.12	0.07	0.19	0.05	0.05	0.08	0.10	0.49
MnO	0.14	0.09	0.23	0.22	tr.	tr.	0.21	0.17	0.14	0.06	0.10	0.10	0.05	0.19	0.21
BaO															
CO_2															0.29
Ba															
Sr															
Rb															
Zr															
Y															
La															
Ce															
Nd															
Sm															
Nb															
Li															
B															
U															
Th															
Mo															
Pb															
Zn															
Cu															
Ni															
Cr															
V															
Co															
$^{87}Sr/^{86}Sr$															
$^{143}Ne/^{144}Nd$															
AGE															
Reference	1	1	1	1	1	1	3	3	3	3	3	3	3	3	4

APPENDIX 7 (continued)

	Locality 5 P'aiktu-san										Locality 6 Longwan				Locality 7 Guandian
SiO_2	61.68	64.42	64.78	65.04	64.89	67.84	70.95	49.05	47.74	46.40	46.74	48.15	48.75	47.52	48.31
Al_2O_3	17.36	16.26	15.00	16.66	15.42	15.27	12.23	14.35	15.41	14.32	14.64	17.49	17.69	17.11	15.10
Fe_2O_3	1.79	2.26	3.73	2.36	2.94	2.22	1.22	3.02	3.96	3.80	4.63	4.14	5.40	1.81	3.60
FeO	4.03	3.77	2.55	2.93	3.68	2.38	4.13	10.57	8.24	7.97	9.48	7.08	5.71	9.48	8.62
MgO	0.37	0.17	0.13	0.02	0.12	0.04	0.05	4.25	8.32	10.15	8.15	7.47	6.85	7.97	6.07
CaO	1.53	1.24	1.21	0.99	1.23	0.82	0.47	8.66	8.33	8.68	8.01	7.21	6.86	7.84	7.16
Na_2O	4.68	5.44	6.28	6.08	5.46	5.32	5.46	3.23	3.82	4.16	3.69	3.23	3.18	3.45	4.67
K_2O	5.10	4.74	4.90	4.88	4.74	5.02	4.60	1.40	2.31	2.79	2.31	2.16	2.40	2.08	3.04
H_2O+	0.27	0.10						0.34			0.49	0.45			
H_2O-	0.39	0.22	0.20	0.14	0.12	0.14	0.04	0.20				0.15	0.81	0.48	
TiO_2	0.65	0.55	0.50	0.40	0.40	0.35	0.30	3.19	1.63	1.75	1.82	2.15	2.04	2.01	2.36
P_2O_5	0.03	0.02	tr.	tr.	tr.		tr.	0.57	0.26		0.50	0.46	0.44	0.38	0.89
MnO	0.13	0.12	0.07	0.09	0.12	0.05	0.09	0.27	0.16		0.22	0.23	0.24	0.22	0.19
BaO															
CO_2	1.80	0.28	0.40	0.16	0.58	0.22	0.36	0.73							
Ba															
Sr															
Rb															
Zr															
Y															
La															
Ce															
Nd															
Sm															
Nb															
Li															
B															
U															
Th															
Mo															
Pb															
Zn															
Cu															
Ni															
Cr															
V															
Co															
$^{87}Sr/^{86}Sr$															
$^{143}Ne/^{144}Nd$															
AGE											H	H	H	H	Q
Reference	4	4	4	4	4	4	4	4	5	5	3	4	4	4	5

APPENDIX 7 (continued)

	Locality 7 Guandian			Locality Nr.9 Jilin	Locality 10 Qixingshan		Locality 11 Yingomen	Locality 15 Muleng						Locality nr.15
SiO_2	49.31	46.85	49.55	45.72	42.23	45.51	44.98	42.72	42.80	46.56	46.68	48.76	49.04	45.73
Al_2O_3	14.92	15.94	15.34	14.11	12.46	14.96	15.56	14.62	14.01	16.59	15.74	16.13	18.02	15.47
Fe_2O_3	4.60	2.95	2.65	3.74	2.89	2.22	5.15	3.62	4.79	2.85	6.23	2.96	4.26	5.16
FeO	6.85	9.32	8.42	7.85	10.93	9.95	7.30	7.88	7.84	7.61	6.58	7.29	6.12	7.58
MgO	8.51	5.22	7.70	10.00	11.81	8.78	3.31	2.95	5.79	6.68	4.47	6.77	4.98	7.56
CaO	7.76	6.67	8.15	8.55	10.31	9.16	9.20	9.18	8.94	8.02	8.76	8.02	6.88	8.59
Na_2O	3.51	5.70	3.64	4.10	3.91	3.50	5.34	3.70	3.80	3.42	3.36	3.06	4.74	3.90
K_2O	2.18	3.00	1.40	2.75	1.73	1.78	1.29	2.84	3.57	3.24	2.26	2.32	3.30	2.10
H_2O+		0.15	0.30		1.30	1.30	3.77	3.01	4.72	0.52	2.13	1.89	0.73	1.47
H_2O-		0.35	0.15											0.22
TiO_2	1.66	2.70	1.55	1.72	2.00	2.05	2.89	2.40	2.22	2.02	2.28	2.01	1.40	1.49
P_2O_5	0.51	0.85	0.27		0.50	0.58	0.43	0.76	1.43	0.59	0.90	0.68	0.60	0.88
MnO	0.19	0.15	0.15	0.00			0.26	0.25	0.22	0.21	0.26	0.24	0.20	0.18
BaO														
CO_2														
Ba		659	276	372										
Sr		1010	555	1096										
Rb		51.6	24.0	30.2										
Zr		298	183	279										
Y		27.7	23.1	24.9										
La														
Ce		163	82.6	170										
Nd		54.7	28.3	57.2										
Sm														
Nb		75.5	34.3	77.9										
Li														
B														
U														
Th														
Mo														
Pb														
Zn														
Cu														
Ni		66.3	193	340										
Cr		41.1	483	428										
V														
Co		156	147	166										
$^{87}Sr/^{86}Sr$		70388	70468	70349										70404
$^{143}Ne/^{144}Nd$														
AGE	Q	Q	Q	Q	H	H								Q
Reference	5	6	6	6	3	3	3	3	3	3	3	3	3	6

APPENDIX 7 (continued)

	Locality 19 Erkeshan				Locality 20 Wudalianchi										
SiO_2	50.58	51.12	51.26	50.99	47.15	49.96	48.70	50.54	50.81	52.68	52.94	53.46	50.76	48.23	50.24
Al_2O_3	14.14	13.76	14.66	14.19	14.27	13.24	17.84	14.50	14.71	14.38	14.69	14.61	14.37	16.25	14.57
Fe_2O_3	4.48	1.28	7.84	4.53	9.32	2.93	3.98	2.31	2.36	1.43	0.89	1.78	2.96	4.54	3.87
FeO	4.61	6.70	1.30	4.20	1.44	7.87	3.87	6.96	6.20	7.76	8.05	3.89	6.10	4.12	4.55
MgO	6.68	6.95	4.21	5.95	7.78	7.48	5.56	7.41	7.43	6.38	6.08	6.16	7.02	6.22	5.34
CaO	5.08	6.44	6.27	6.15	8.32	8.02	6.04	7.01	6.65	5.82	6.01	5.92	6.60	6.13	5.66
Na_2O	4.08	4.22	4.54	4.22	3.25	3.79	3.45	3.57	3.56	3.36	3.35	3.33	3.72	4.52	3.69
K_2O	4.01	5.61	5.27	5.61	4.56	4.89	5.31	4.28	5.61	5.43	5.13	5.20	4.76	4.48	5.01
H_2O+	1.26	2.61	1.19		0.79	0.42	1.71	0.69	0.29	0.11	0.32	0.37			
H_2O-				0.98									0.51		
TiO_2	2.56	2.45	2.42	2.48	2.46	2.83	2.58	2.05	2.19	1.88	2.00	2.77	2.38	2.50	2.52
P_2O_5	1.04	1.06	1.14	1.08	1.07	1.03	1.21	1.05	1.06	1.10	1.01	0.92	1.04	1.55	1.24
MnO	0.10	0.12	0.08	0.10	0.18	0.02	0.12	0.55	0.14	0.15	0.15	0.01	0.11		0.09
BaO															
CO_2															
Ba															
Sr															
Rb															
Zr															
Y															
La															
Ce															
Nd															
Sm															
Nb															
Li															
B															
U															
Th															
Mo															
Pb															
Zn															
Cu															
Ni															
Cr															
V															
Co															
$^{87}Sr/^{86}Sr$															
$^{143}Ne/^{144}Nd$															
AGE	H	H	H	H										PRE-H	PRE-H
Reference	3	3	3	4	3	3	3	3	3	3	3	3	4	7	7

APPENDIX 7 (continued)

	Locality 20 Wudalianchi								Locality 24 Nr.Nenjiang			Locality SSE of 27 Lubei	Locality 28 Daerbo
SiO_2	53.90	53.96	53.96	50.92	54.50	54.36	52.46	53.28	42.84	52.12	53.36	47.31	42.98
Al_2O_3	13.90	13.93	13.70	13.37	14.50	13.81	14.26	15.50	12.13	13.89	13.15	15.24	13.67
Fe_2O_3	2.94	2.94	2.30	2.82	3.36	8.14	5.39	6.76	5.86	3.79	1.67	10.70	4.13
FeO	5.57	5.57	6.01	5.93	4.91	0.41	3.48	2.78	4.43	4.65	6.37	4.95	9.45
MgO	5.65	5.88	5.91	7.90	4.90	4.67	6.69	5.36	11.80	6.56	6.13	7.26	7.89
CaO	5.66	5.45	5.40	6.55	5.50	5.55	6.07	5.85	10.54	6.40	5.47	8.69	10.06
Na_2O	3.67	3.79	3.67	3.92	3.88	3.78	3.96	3.22	2.37	3.40	3.55	2.98	4.27
K_2O	5.22	5.06	5.22	4.81	5.47	5.47	5.40	5.61	3.68	4.94	5.76	1.45	2.11
H_2O+									3.44	0.92	0.92		0.68
H_2O-													
TiO_2	2.40	2.20	2.20	2.20	2.54	2.40	2.42	2.02	2.49	2.57	2.57	0.25	3.18
P_2O_5	0.99	0.98	1.02	0.95	1.15	1.01	0.33	0.24	0.55	0.71	0.68	1.11	0.87
MnO	0.11	0.11	0.11	0.10	0.11	0.11	0.06	0.06	0.17	0.25	0.17	0.10	0.21
BaO													
CO_2													
Ba													
Sr													
Rb													
Zr													
Y													
La													
Ce													
Nd													
Sm													
Nb													
Li													
B													
U													
Th													
Mo													
Pb													
Zn													
Cu													
Ni													
Cr													
V													
Co													
$^{87}Sr/^{86}Sr$													
$^{143}Ne/^{144}Nd$													
AGE	HIST	HIST	HIST	HIST	HIST	HIST	HIST	HIST				ET	
Reference	7	7	7	7	7	7	7	7	3	3	3	5	3

APPENDIX 7 (continued)

	Locality 28 Daerbo	Locality 30 Weizhang			Locality 32-36 Hebei-Shansi-Beijing									
SiO_2	47.06	45.42	46.12	47.88	49.86	50.03	49.00	50.09	48.95	49.00	49.77	49.22	50.78	48.84
Al_2O_3	13.81	13.10	16.38	15.09	14.76	14.15	13.61	14.15	14.97	14.16	14.80	15.35	14.53	15.84
Fe_2O_3	3.27	4.51	3.72	3.59	10.30	3.55	3.71	3.60	3.79	11.52	12.41	11.77	2.43	12.44
FeO	8.64	7.66	6.63	8.59		7.45	7.61	7.53	7.79				7.98	
MgO	9.21	9.30	3.26	4.96	8.43	7.12	7.23	6.51	5.57	9.90	7.44	6.43	7.60	7.20
CaO	9.82	10.62	10.16	8.82	8.10	8.01	7.41	8.49	9.01	8.05	8.12	7.86	7.80	8.04
Na_2O	3.27	2.82	4.67	3.31	2.92	2.15	2.20	2.64	2.80	2.58	2.87	3.42	3.25	3.11
K_2O	1.34	1.57	3.07	1.59	1.82	1.10	0.95	0.54	0.54	0.28	0.90	1.32	0.93	1.42
H_2O+	0.94	1.19	1.35	2.92	1.96	2.40	3.06	2.35	1.63	2.70	2.50	2.74	1.13	2.00
H_2O-			0.56			2.35	3.95	2.25	2.45					
TiO_2	2.22	2.42	3.04	2.52	1.97	1.58	1.35	1.56	1.86	2.17	1.76	2.19	2.00	1.69
P_2O_5	0.49	0.50	0.81	0.58	0.53	0.40	0.32	0.32	0.48	0.34	0.37	0.65	0.68	0.55
MnO	0.17	0.23	0.18	0.19	0.15	0.21	0.18	0.18	0.24	0.20	0.12	0.14	0.14	0.14
BaO														
CO_2														
Ba					368	439	749	2789	248	189	246	1096	248	468
Sr					435	560	566	527	505	508	517	1154	504	607
Rb					41.7	15.8	8.1	1.5	2.2	1.9	13.6	22.0	15.6	12.7
Zr					231	141	99.1	120	118	118	128	147	132	175
Y					30.1	23.3	18.8	23	23	26	22.8	27.5	21.0	25.5
La														
Ce					93.9	57.0	20.3	38.9	51.8	36.8	40.0	60.9	53.3	61.2
Nd					35.8	24.0	12.0	18.9	17.1	19.3	18.3	24.9	17.7	25.7
Sm														
Nb					40.6	24.7	14.9	19.4	16.6	14.2	18.4	28.0	24.3	23.0
Li														
B														
U														
Th														
Mo														
Pb														
Zn														
Cu														
Ni					246	251	180	155	204	155	18	240	137	118
Cr						212	147	124	175	158	166	200	152	200
V							157	176	179	205	182	206	162	176
Co						186								
$^{87}Sr/^{86}Sr$					70337	70420	70457	70436	70454	70428	70427	70398	70483	70483
$^{143}Nd/^{144}Nd$														
AGE					ET	ET	ET	ET	ET	ET	ET	ET	N	N
Reference	3	3	3	3	6	6	6	6	6	6	6	6	6	6

APPENDIX 7 (continued)

	Locality 32-36 Hebei-Shansi-Bejing														
SiO_2	48.86	49.32	50.67	50.87	44.97	42.76	46.74	36.03	48.96	49.50	49.99	48.16	47.23	47.92	50.95
Al_2O_3	15.89	15.85	13.11	14.79	14.45	14.29	13.48	9.87	14.73	15.55	17.20	14.74	14.67	14.53	15.98
Fe_2O_3	12.21	11.19	2.82	0.75	12.86	15.42	4.91	4.69	12.71	13.80	12.13	2.75	3.87	5.37	5.09
FeO			9.88	10.94			9.22	10.72				7.45	8.91	7.49	8.05
MgO	6.65	6.48	8.00	6.59	8.90	6.61	6.63	8.15	5.03	4.43	4.77	6.55	7.78	8.19	5.48
CaO	7.99	8.71	8.16	7.90	7.26	7.79	6.67	9.88	7.13	6.86	8.60	10.06	8.51	8.51	8.63
Na_2O	3.15	2.40	3.00	2.86	4.80	5.31	4.06	4.40	5.24	4.92	4.26	3.28	3.89	3.08	2.47
K_2O	1.64	1.83	0.80	0.60	2.03	2.58	2.64	2.40	1.72	1.70	0.79	2.09	2.12	1.61	1.06
H_2O+	2.23	2.50	1.20	0.96	2.20	2.76	1.64		2.00	1.50		2.42			
H_2O-				0.61			0.31	0.95							
TiO_2	1.99	1.88	2.11	1.95	2.52	2.59	2.69	2.39	2.45	2.33	1.91	2.00	2.35	2.31	1.96
P_2O_5	0.54	0.81	0.38	0.20	1.11	1.38	0.60	1.42	0.99	0.72	0.21	0.14	1.05	0.57	0.34
MnO	0.15	0.13	0.18	0.18	0.15	0.17	0.17	0.20	0.16	0.14		0.16	0.21	0.17	0.09
BaO															
CO_2															
Ba	357	518	237	187	676	451	611	834	398	352					
Sr	528	1223	371	330	1333	1229	1050	1798	732	715	413				
Rb	17.5	25.4	7.85	10.9	46.4	36.9	60.8	41.8	27.8	26.0	13.3				
Zr	152	136	130	140	270	263	402	255	290	270	158				
Y	24.3	22.6	25.1	29.3	23.6	25	34.2	29.9	35.1	34.8	28.2				
La															
Ce	61.8	93.5	35.0	33.9	119	110	156	141	103	94.4					
Nd	20.8	28.7	15.9	14.6	44.3	43.9	59.0	57.1	41.6	37.9					
Sm															
Nb	35.5	37.9	18.8	14.7	74.6	80.5	80.4	103	51.4	46.1	19.7				
Li															
B															
U															
Th															
Mo															
Pb															
Zn															
Cu															
Ni	210	94.2	190	148	161	161	191	329	194	154					
Cr	195	172	193	156	115	168	242	374	192	178					
V	174	165	191	205	182	187	191	159	185	163					
Co															
$^{87}Sr/^{86}Sr$	70441	70464	70398	70427	70397	70383	70404	70395	70351	70355	70351				
$^{143}Nd/^{144}Nd$															
AGE	N	N	N	N	N	N	N?	N?	Q	Q	Q		LT	LT	LT
Reference	6	6	6	6	6	6	6	6	6	6	6	3	5	5	5

APPENDIX 7 (continued)

	Locality 32-36 Hebei-Shansi-Beijing									Locality 37-38 S. Heibei-Shansi-N. Henan					
SiO_2	48.67	52.32	51.52	51.34	46.24	47.64	45.51	48.21	46.13	47.52	42.48	48.12	49.06	47.26	48.44
Al_2O_3	15.38	16.25	16.36	15.64	16.19	15.17	15.27	14.72	14.62	14.71	12.33	15.45	15.13	14.39	14.38
Fe_2O_3	6.75	3.42	3.94	4.44	4.46	8.57	3.29	5.49	7.14	7.22	4.99	12.74	11.08	13.00	4.37
FeO	5.70	6.95	8.11	7.34	6.66	2.85	9.62	7.87	6.40	4.58	9.01				6.60
MgO	7.45	6.91	5.24	6.95	7.27	7.63	8.05	6.47	6.36	4.98	9.41	5.91	6.76	6.82	8.95
CaO	9.51	7.82	8.66	8.53	9.08	10.08	7.79	7.79	8.17	9.00	11.34	7.79	7.84	7.49	7.75
Na_2O	3.09	2.83	2.79	2.66	3.36	3.28	4.20	3.82	3.34	5.18	4.46	4.18	4.70	4.16	3.07
K_2O	1.10	0.98	0.71	0.74	1.66	1.34	2.25	1.85	3.77	1.20	1.84	2.26	1.44	2.30	1.65
H_2O+					1.88	0.72						1.50	1.50	2.00	
H_2O-															0.07
TiO_2	1.82	1.91	2.01	1.76	2.02	2.02	2.82	2.78	3.05	3.53	3.08	2.48	2.46	2.58	2.85
P_2O_5	0.31	0.42	0.47	0.37	0.51	0.65	0.90	0.80	0.86	1.92	0.95	0.76	0.93	0.98	0.71
MnO	0.21	0.17	0.19	0.19	0.12	0.18	0.30	0.20	0.16	0.16	0.12	0.14	0.13	0.15	0.16
BaO															
CO_2															
Ba												468	374	421	455
Sr												847	885	990	812
Rb												27.9	20.6	28.2	18.3
Zr												211	249	284	167
Y												26.0	28.7	30.9	23.1
La															
Ce												78.6	99.0	108	65.4
Nd												35.7	36.9	45.8	30.4
Sm															
Nb												56.7	59.9	74.6	48.6
Li															
B															
U															
Th															
Mo															
Pb															
Zn															
Cu															
Ni												151	155	174	223
Cr												165	129	214	
V												159	166	169	
Co															
$^{87}Sr/^{86}Sr$												70355	70332	70406	70454
$^{143}Nd/^{144}Nd$															
AGE	ET	ET	ET	ET			Q	Q	Q	Q	Q	Q	Q	Q	N?
Reference	5	5	5	5	3	5	5	5	5	5	5	6	6	6	6

APPENDIX 7 (continued)

	Locality 39 Shangdong	Locality 40 North Bohai						Locality 40 Jingyuan	Locality 41 NE. Shangdong					Locality 42-43 around Nanjing	
SiO_2	45.88	43.51	46.33	45.55	44.29	49.74	50.43	47.18	41.13	42.64	39.75	41.03	41.38	45.08	48.66
Al_2O_3	14.90	14.27	16.77	13.87	15.37	17.70	19.60	16.21	12.00	12.32	11.20	11.71	12.09	14.55	13.95
Fe_2O_3	14.35	11.26	10.75	11.53	10.03	5.40	5.59	3.26	4.27	5.56	5.64	3.41	5.84	9.20	4.40
FeO						4.17	5.95	9.99	9.94	7.85	8.55	9.62	9.25	4.11	8.04
MgO	8.65	11.01	7.43	10.47	10.93	5.34	4.51	7.20	9.42	9.14	8.52	10.70	8.43	9.32	8.45
CaO	8.38	9.21	9.09	7.57	7.17	9.52	6.29	8.46	10.97	11.38	8.52	10.10	8.99	8.54	8.81
Na_2O	3.19	1.50	2.58	1.96	2.92	4.40	3.15	3.28	5.22	3.58	3.94	4.65	5.00	4.39	3.31
K_2O	1.72	1.07	1.35	1.50	1.20	1.43	2.22	1.62	2.24	1.39	1.80	2.00	2.00	1.13	1.52
H_2O+	1.70								0.68	2.53		2.01	2.73		
H_2O-									0.27		1.32				
TiO_2	2.12	2.21	2.03	1.82	2.07	1.65	1.75	2.08	2.62	3.44	2.64	2.28	2.44	2.21	2.22
P_2O_5	0.55	0.50	0.54	0.40	0.68	0.48	0.43	0.63	1.32	0.57	0.75	1.51	0.96	1.29	0.67
MnO	0.16	0.25	0.18	0.10	0.15	0.16	0.15	0.21	0.17	0.06	0.20	0.22	0.15	0.19	0.16
BaO															
CO_2															
Ba	224	448	217	484	318						990	214	687		
Sr	571	811	630	607	695						1553	1252	1672		
Rb	20.8	12.1	7.43	18.2	8.56						17.5	14.9	35.4		
Zr	157	150	206	169	197						281	308	339		
Y	20.6	23.3	30.0	22.9	27.2						32.2	31.6	40.7		
La															
Ce	50.3	52.1	82.9	84.3	86.5						205	202	227		
Nd	21.5	21.7	29.5	29.4	30.8						66.9	67.1	81.1		
Sm															
Nb	33.8	30.1	33.4	34.0	43.3						114	119	142		
Li															
B															
U															
Th															
Mo															
Pb															
Zn															
Cu															
Ni	296	237	55.1	201	174						179	169	146		
Cr		311	60		208						180	165	176		
V		217	185		162						181	175	130		
Co															
$^{87}Sr/^{86}Sr$	70306	70351	70314	70385	70316						70386	70378	70328		
$^{143}Nd/^{144}Nd$															
AGE	N	ET	ET	ET	ET	ET	ET	LT			N	N	N	ET	ET
Reference	6	6	6	6	6	5	5	5	3	3	6	6	6	5	5

APPENDIX 7 (continued)

	Locality 42-43 around Nanjing															
SiO_2	51.17	51.70	47.61	52.31	46.55	45.48	46.72	53.71	49.12	47.35	42.75	46.82	46.74	44.25	44.82	47.17
Al_2O_3	15.78	13.92	14.25	15.76	14.51	15.10	13.01	16.29	15.35	13.77	13.06	13.73	13.86	15.23	17.59	15.16
Fe_2O_3	7.43	11.72	10.95	10.81	13.29	12.13	11.91	9.22	10.78	12.70	13.80	12.41	4.06	7.71	6.32	4.03
FeO	4.05												9.60	3.42	5.21	6.88
MgO	5.63	7.80	5.98	6.56	9.01	6.66	6.57	3.96	6.54	9.57	9.79	9.42	6.44	4.53	5.78	7.39
CaO	9.04	7.99	9.97	7.15	7.10	8.39	7.83	7.62	8.45	8.28	9.68	8.17	7.59	4.02	8.51	8.41
Na_2O	3.25	2.82	3.20	3.75	3.97	4.61	6.49	4.88	4.20	2.58	6.04	3.43	4.52	1.27	2.16	3.21
K_2O	1.36	0.21	0.92	1.54	1.41	1.79	2.68	1.69	1.21	1.64	1.43	1.83	3.21	0.75	1.21	1.07
H_2O+				1.02	2.19	3.59	2.00	1.50	2.96	2.00	2.00	1.50		3.96	2.83	1.28
H_2O-														8.23	3.56	1.57
TiO_2	2.04	1.37	1.36	1.48	2.05	2.35	2.10	1.33	1.83	2.29	1.92	2.27	2.54	1.61	1.70	3.12
P_2O_5	0.41	0.19	0.39	0.37	0.50	1.02	1.36	0.48	0.44	0.67	0.86	1.16	1.27	0.82	0.73	0.69
MnO	0.18	0.14	0.11	0.13	0.15	0.14	0.16	0.11	0.14	0.15	0.17	0.16	0.18	0.09		
BaO																0.22
CO_2																
Ba		109	314	253	354	441	581	307	335	406	594	543				
Sr		294	576	452	689	1084	1234	491	601	857	863	895				
Rb		1.22	18.0	22.3	49.1	39.7	44.3	29.7	36.4	25.9	42.3	26.2				
Zr		108	121	112	166	212	255	124	129	157	163	209				
Y		21.44	23.0	19.2	22.5	25.0	36.4	22.6	24.1	26.2	25.7	29.7				
La																
Ce		22.3	34.3	27.8	58.7	87.9	165	29.4	36.1	86.7	59.9	81.6				
Nd		14.0	17.2	15.2	25.7	36.1	52.4	15.2	16.2	29.5	28.2	36.2				
Sm																
Nb		10.0	23.2	27.1	49.9	69.8	96.8	29.0	30.7	44.7	73.5	55.9				
Li																
B																
U																
Th																
Mo																
Pb																
Zn																
Cu																
Ni		217	146	194	370	138	143	99.2	146	291	306	210				
Cr		238	280	176	390		227	136	194	361		185				
V		167	185	133	160		135	149	175	180		176				
Co																
$^{87}Sr/^{86}Sr$		70444	70396	70368	70367	70361	70388	70399	70421	70406	70339	70377				
$^{143}Nd/^{144}Nd$																
AGE	ET	ET	ET	N	N	N	N	N	N	N	N	N	Q			
Reference	5	6	6	6	6	6	6	6	6	6	6	6	5	3	3	3

APPENDIX 7 (continued)

	Locality 42-43 around Nanjing															
SiO_2	47.[illegible]0	48.26	49.95	50.10	50.89	51.86	46.28	49.75	51.04	48.59	48.92	51.86	48.08	51.26	51.80	51.62
Al_2O_3	16.[illegible]8	15.27	16.30	17.78	18.33	14.80	11.84	16.25	17.12	8.67	18.19	8.83	14.99	14.48	15.71	15.67
Fe_2O_3	5.[illegible]6	6.51	8.87	6.38	7.78	7.83	9.04	5.95	6.10	9.93	8.03	10.97	6.02	4.47	4.44	3.71
FeO	5.[illegible]6	4.49	2.72	4.09	2.57	2.72	4.44	4.81	3.82	6.41	3.33	5.30	4.89	6.39	6.66	7.61
MgO	5.[illegible]1	6.21	4.12	4.47	3.65	4.11	6.58	5.55	4.72	7.56	2.59	4.57	7.07	5.52	5.18	5.38
CaO	8.[illegible]1	8.34	9.14	8.07	7.38	8.06	8.97	8.55	8.13	8.59	6.41	8.90	8.58	9.28	8.60	9.65
Na_2O	2.[illegible]1	4.32	3.89	3.86	3.86	3.81	4.47	3.83	4.56	3.45	4.49	4.67	3.66	3.16	3.47	3.19
K_2O	1.[illegible]3	1.14	1.55	1.76	1.91	1.97	1.21	1.45	1.66	1.19	1.70	1.59	2.11	1.21	0.50	0.72
H_2O+	0.[illegible]5	0.86	0.19	0.35	0.94	1.10	1.76	1.45	1.05	2.68	1.49	0.67	0.50	0.90	1.15	
H_2O-	1.[illegible]3	0.47	0.81	1.25	1.23	1.30	1.92	1.39	1.20	1.67	3.96	0.83	1.10	1.20	1.32	
TiO_2	1.[illegible]5	3.57	2.58	1.53	1.53	2.25	3.47	1.19	1.22	1.95	1.45	2.35	2.13	2.13	1.31	1.92
P_2O_5	1.[illegible]3	0.07	0.58	0.46	0.43	0.51	0.43	0.23	0.26	0.22	0.33	0.28		0.44	0.25	0.29
MnO	0.[illegible]9	0.12	0.13	0.10			0.05	0.07	0.08	0.01	0.07	0.04	0.52		0.01	0.26
BaO																
CO_2																
Ba																
Sr																
Rb																
Zr																
Y																
La																
Ce																
Nd																
Sm																
Nb																
Li																
B																
U																
Th																
Mo																
Pb																
Zn																
Cu																
Ni																
Cr																
V																
Co																
$^{87}Sr/^{86}Sr$																
$^{143}Nd/^{144}Nd$																
AGE																ET
Reference	3	3	3	3	3	3	3	3	3	3	3	3	3	3	3	5

APPENDIX 7 (continued)

	Locality 42-43 around Nanjing		South of Locality 45 Shantou	Locality 46 Leizhou Bendao	Locality 47 Hainan Dao			Locality 48 Weizhou Dao	Locality 49 Tengchong					
SiO_2	49.48	54.72	43.33	51.79	53.32	53.71	48.64	48.98	54.00	67.98	67.99	67.42	53.60	52.04
Al_2O_3	13.94	17.90	14.19	15.60	15.79	16.08	14.22	14.12	17.44	14.65	15.56	14.57	17.50	18.66
Fe_2O_3	3.76	2.99	11.67	10.81	12.17	10.62	12.29	2.25	9.51	3.66	1.69	0.57		
FeO	7.48	5.61						7.40			1.72	4.35	8.73	8.84
MgO	9.04	5.64	17.76	6.87	3.80	5.39	7.98	7.98	3.49	1.21	1.50	1.60	5.85	3.99
CaO	10.81	5.96	11.08	5.40	8.52	7.85	6.99	10.30	7.82	3.82	2.48	3.19	8.08	10.12
Na_2O	2.98	2.34	0.30	3.81	3.46	3.60	4.03	2.61	4.24	2.96	3.00	3.00	2.74	3.09
K_2O	0.61	1.50	0.18	2.03	0.43	0.71	1.61	1.61	1.92	3.55	4.30	3.80	1.52	1.21
H_2O+			1.50	1.50	1.50		1.30	1.02	0.55	0.50	0.12			
H_2O-								0.32						
TiO_2	1.37	2.63	1.15	2.04	1.69	1.62	2.95	2.88	1.59	0.59	0.54	0.61	1.44	1.55
P_2O_5	0.28	0.47	0.05	0.86	0.14	0.21	0.87	0.30	0.38	0.14	0.14	0.33	0.45	0.15
MnO	0.25	0.21	0.13	0.10	0.25	0.15		0.14	0.14	0.07	0.06	0.07	0.15	0.14
BaO														
CO_2														
Ba			401	365	69.0		586		447	584	615	606		
Sr			31.6	726	249	299	900		485	336	315	321		
Rb			6.45	36.1	8.54	13.8	56.9		38.8	163	167	164		
Zr			52.1	223	104	122	272		161	219	232	258		
Y			16.6	17.7	48.4	21.2	29.0		28.0	20.4	23.1	22.8		
La														
Ce			≤10	83.2	15.8		111		60.9	121				
Nd			≤3.0	34.2	12.0		46.7		25.3	47.8				
Sm														
Nb			5.38	47.3	8.50	14.0	67.3		17.3	15.7	17.4	16.3		
Li														
B														
U														
Th														
Mo														
Pb														
Zn														
Cu														
Ni			171	87.0	87	142	212		46.4	8.8	19.9	25.0		
Cr			36	65.3	131				114			59.4		
V			360	117	154				153			83.0		
Co														
$^{87}Sr/^{86}Sr$			70310	70309	70345	70366	70295		70656	70908	70930	70908		
$^{143}Nd/^{144}Nd$														
AGE	ET	ET	Q	Q	Q	Q	Q		Q	Q	Q	Q	E.PLI	E.PLI
Reference	5	5	6	6	6	6	6	3	6	6	6	6	8	8

APPENDIX 7 (continued)

	Locality 49 Tengchong															
SiO_2	53.47	52.96	50.16	49.41	49.92	49.30	51.56	64.23	67.27	66.00	66.63	68.99	67.48	62.90	53.75	54.58
Al_2O_3	17.78	19.38	16.81	16.31	16.62	15.63	16.68	16.67	14.99	15.00	14.74	16.12	14.89	16.65	17.96	17.76
Fe_2O_3																
FeO	7.39	6.91	10.53	9.93	9.18	9.48	10.26	3.99	3.49	3.70	3.74	3.23	3.69	4.50	7.49	7.80
MgO	5.03	4.19	6.46	6.79	8.07	8.09	6.78	1.95	1.80	3.51	1.81	1.27	1.36	1.87	5.12	4.73
CaO	7.38	7.75	8.67	9.32	7.93	8.95	8.07	4.25	2.69	3.62	3.61	3.85	2.86	4.65	5.57	7.25
Na_2O	3.32	3.25	3.33	3.46	2.85	2.66	3.45	3.12	3.07	2.90	3.39	2.74	3.11	3.35	3.25	3.84
K_2O	2.08	1.85	1.01	1.19	1.63	1.45	1.36	3.15	3.85	3.82	3.90	3.27	4.09	3.34	2.77	2.50
H_2O+																
H_2O-																
TiO_2	1.30	1.18	1.53	1.39	1.39	1.30	1.31	0.76	0.54	0.45	0.51	0.51	0.50	0.87	1.24	1.31
P_2O_5	0.37	0.42	0.23	0.34	0.47	0.55	0.27	0.30	0.16	0.21	0.18	0.20	0.32	0.32	0.36	0.32
MnO	0.15	0.12	0.17	0.16	0.17	0.22	0.19	0.11	0.07	0.12	0.13	0.08	0.12	0.07	0.16	0.14
BaO																
CO_2																
Ba			248		645	936	442		489			538	767			
Sr			396		588	897	431		267			318	379			
Rb			16.4		38.5	27.1	27.0		198			164	133			
Zr			137		173	185	192		222			268	300			
Y																
La																
Ce																
Nd			18.4		30.0	44.5	23.7		60.8			46.4	47.7			
Sm			4.13		5.29	7.04	4.66		9.60			7.29	8.11			
Nb																
Li																
B																
U																
Th																
Mo																
Pb																
Zn																
Cu																
Ni			71.0		110	113	112		20.0			4.6	19.4			
Cr																
V			162		138	138	146		47.3			53.6	64.9			
Co																
$^{87}Sr/^{86}Sr$			70595		70714	70803	70691		71124			70925	70813			
$^{143}Nd/^{144}Nd$			51260		51246	51245	51249		51193			51203	51223			
AGE	E.PLI	E.PLI	E.PLI	E.PLI	L.PLI	L.PLI	L.PLI	E.PLE	E.PLE	E.PLE	E.PLE	E.PLE	E.PLE	E.PLE	L.PLE	L.PLE
Reference	8	8	8	8	8	8	8	8	8	8	8	8	8	8	8	8

APPENDIX 7 (continued)

	Locality 49 Tengchong														
SiO_2	54.58	54.56	55.43	51.24	55.03	53.54	55.09	53.94	50.32	52.68	57.59	57.88	57.18	60.03	62.48
Al_2O_3	17.63	17.50	15.83	18.06	16.88	17.96	17.70	17.27	17.81	17.16	17.17	16.01	16.98	17.47	16.41
Fe_2O_3															
FeO	8.10	8.23	7.84	8.59	7.39	8.73	7.39	8.55	8.76	8.65	6.34	6.46	6.64	5.86	5.46
MgO	5.07	3.31	5.23	5.65	4.28	3.68	3.58	5.28	6.82	5.26	3.87	4.09	4.04	2.93	2.34
CaO	7.56	8.75	6.78	8.87	8.27	8.33	6.25	7.92	8.45	7.39	6.00	5.52	5.88	5.64	5.33
Na_2O	3.39	2.90	3.38	3.12	2.80	2.98	3.62	3.68	2.90	3.28	3.48	3.79	3.50	3.22	2.90
K_2O	1.89	2.16	2.48	1.76	2.20	2.09	2.58	1.81	1.57	1.69	2.76	3.61	3.09	3.17	3.86
H_2O+															
H_2O-															
TiO_2	1.41	1.38	1.19	1.45	1.27	1.21	1.39	1.44	1.32	1.37	1.18	1.11	1.24	1.19	0.98
P_2O_5	0.20	0.44	0.29	0.56	0.42	0.50	0.50	0.28	0.47	0.47	0.41	0.41	0.56	0.28	0.42
MnO	0.12	0.20	0.13	0.20	0.17	0.18	0.16	0.14	0.20	0.14	0.12	0.15	0.18	0.10	0.16
BaO															
CO_2															
Ba	521					758		384						873	
Sr	451					874		465						580	
Rb	45.1					70.0		43.9						98.9	
Zr	230					233		200						368	
Y															
La															
Ce															
Nd	33.2					53.6		26.6						60.0	
Sm	6.11					8.31		5.24						9.55	
Nb															
Li															
B															
U															
Th															
Mo															
Pb															
Zn															
Cu															
Ni	40.5					36.1		45.4						35.6	
Cr															
V	141					154		146						105	
Co															
$^{87}Sr/^{86}Sr$	70672					70595		70645						70880	
$^{143}Nd/^{144}Nd$	51234					51236		51244						51222	
AGE	L.PLE	L.PLE	L.PLE	L.PLE	L.PLE	L.PLE	L.PLE	L.PLE	L.PLE	L.PLE	H	H	H	H	H
Reference	8	8	8	8	8	8	8	8	8	8	8	8	8	8	8

APPENDIX 7 (continued)

	Locality 49 Tengchong				Locality 50 Bamaoqiongzhong						Locality 50 Yongbohu		Locality 50 Qiangbaqian		
SiO_2	62.50	59.34	61.16	60.07	44.90	48.97	52.68	56.35	45.05	43.25	67.54	69.25	62.72	58.53	62.20
Al_2O_3	16.75	17.01	16.25	16.04	12.42	14.89	15.64	18.11	12.78	12.14	14.44	14.53	14.80	14.44	14.37
Fe_2O_3					10.79	6.42	5.64	5.29	6.85	6.31	2.91	2.91	3.14	3.87	6.09
FeO	5.20	5.90	5.46	6.22	2.51	1.90	1.01	0.79	2.84	3.05	0.68	0.25	1.96	2.99	1.09
MgO	2.49	3.89	3.90	2.60	7.55	3.40	2.55	1.16	8.08	7.47	1.16	0.73	1.64	2.19	1.39
CaO	4.80	5.87	4.94	5.55	10.99	7.93	5.06	1.94	10.29	9.29	7.87	1.43	3.37	5.82	3.37
Na_2O	3.61	3.46	2.09	2.65	2.75	4.05	2.90	6.25	3.00	4.05	4.25	4.20	3.90	3.45	4.05
K_2O	3.91	3.15	3.58	3.12	5.25	6.40	7.50	6.45	3.95	2.60	4.30	5.15	4.20	3.75	3.70
H_2O+															
H_2O-															
TiO_2	0.99	1.13	0.93	1.05	1.20	1.27	2.63	0.35	1.20	1.68	0.79	0.79	1.47	1.46	1.62
P_2O_5	0.36	0.34	0.42	0.38	0.18	0.70	1.07	0.04	1.33	1.33	0.37	0.31	1.22	1.27	1.29
MnO	0.11	0.13	0.12	0.15	0.16	0.08	0.08	0.12	0.10	0.12	0.04		0.04	0.08	0.04
BaO															
CO_2															
Ba		918													
Sr		558													
Rb		103													
Zr		362													
Y															
La															
Ce															
Nd		53.6													
Sm		8.69													
Nb															
Li															
B															
U															
Th															
Mo															
Pb															
Zn															
Cu															
Ni		56.8													
Cr															
V		105													
Co															
$^{87}Sr/^{86}Sr$		70851													
$^{143}Nd/^{144}Nd$		51222													
AGE	H	H	H	H	Q	Q	Q	Q	Q	Q	Q	Q	Q	Q	Q
Reference	8	8	8	8	9	9	9	9	9	9	9	9	9	9	9

APPENDIX 7 (continued)

	Locality 50 36.16N, 87.5E	Locality 50 Zhongguo-shankou	Locality 50 Kala-Kashenhegu		Locality 50 Zhixierhe	Locality 50 Ali-nazar qurghan	Locality 54 Zangxibei-Xikunlun			Locality 58 Jade Mines	
SiO_2	61.45	58.18	46.09	46.80	43.17	45.15	55.21	55.24	56.77	55.37	62.00
Al_2O_3	13.36	14.49	13.48	14.30	15.14	15.13	14.83	14.95	15.07	16.90	18.80
Fe_2O_3	2.75	2.17	6.15	8.50	10.79	1.20	5.02	4.87	3.63	5.28	2.96
FeO	4.61	3.44	2.70	1.52	2.58	5.22	2.05	1.39	4.67	1.65	1.04
MgO	2.73	3.90	8.79	7.82	7.35	6.21	4.25	3.55	2.64	4.61	1.00
CaO	4.34	5.89	8.41	8.42	8.64	9.73	6.24	6.18	4.97	5.93	4.71
Na_2O	3.98	3.85	2.17	3.86	2.86	2.29	2.88	2.82	2.88	3.58	4.27
K_2O	3.75	4.95	5.31	2.80	1.48	6.63	4.82	4.44	3.55	2.34	2.31
H_2O+										1.85	1.41
H_2O-										0.92	0.72
TiO_2	1.37	1.62	1.72	1.80	2.47	3.54	2.35	2.18	2.38	0.40	0.41
P_2O_5		1.17	1.04		0.35	1.11	0.89	1.01	0.99	0.33	0.34
MnO		0.05	0.13	0.14	0.13	0.10	0.07	0.08	0.11	0.09	0.15
BaO										0.21	0.07
CO_2											
Ba											
Sr											
Rb											
Zr											
Y											
La											
Ce											
Nd											
Sm											
Nb											
Li											
B											
U											
Th											
Mo											
Pb											
Zn											
Cu											
Ni											
Cr											
V											
Co											
$^{87}Sr/^{86}Sr$											
$^{143}Nd/^{144}Nd$											
AGE	Q?	Q?	Q?	Q?	Q?	Q?	Q?	Q?	Q?	LC	LC
Reference	9	9	9	9	9	9	9	9	9	10	10

APPENDIX 7 (continued)

	Locality 59 Taungthonlon	Locality 60 Tigyaing		Locality 61 Singu	Locality 63 Kyaukmyet	Locality 63 Sonetaung		Locality 63 Twindaung		Locality 63 Taungbyauk	Locality 63 Twinywa	Locality 63 Twindaung
SiO_2	56.21	50.18	46.21	51.69	74.95	49.44	46.99	47.69	56.98	47.28	44.66	51.19
Al_2O_3	17.52	15.87	15.71	17.17	12.04	14.40	14.59	18.20	18.24	15.28	15.14	17.72
Fe_2O_3	4.54	2.07	2.64	2.09	0.24	3.34	4.14	5.26	2.82	5.58	6.03	7.61
FeO	2.72	6.60	7.34	6.15	2.54	5.45	5.45	4.18	3.55	3.93	2.50	
MgO	4.02	7.73	8.16	5.53	0.77	10.00	8.03	6.02	3.98	9.43	10.21	5.20
CaO	5.90	7.87	10.15	6.62	0.50	10.79	13.20	11.57	8.48	12.77	13.38	10.19
Na_2O	5.32	3.93	4.35	3.91	3.11	2.29	3.52	2.74	2.69	2.36	4.29	3.36
K_2O	1.18	2.47	1.54	3.15	3.66	1.33	1.16	1.39	2.55	1.23	1.13	2.54
H_2O+				1.07	2.08			1.24	0.33	0.61	0.32	
H_2O-				0.13	0.14			0.21		0.33	0.04	
TiO_2	0.08	0.25	0.25	2.13	0.25	0.95	0.17	0.80	0.47	0.49	1.85	0.77
P_2O_5	0.10	0.21	0.25	0.63		0.22	0.11	0.36		0.42	0.05	0.40
MnO	0.10	0.11	0.13	0.11		0.16	0.12	0.17	0.23	0.18	0.09	0.12
BaO								0.08		0.05		
CO_2												
Ba												1517
Sr												1567
Rb												40
Zr												145
Y												20
La												127
Ce												142
Nd												
Sm												0
Nb												
Li												
B												
U												
Th												25
Mo												
Pb												30
Zn												53
Cu												235
Ni												52
Cr												61
V												217
Co												51
$^{87}Sr/^{86}Sr$												28
$^{143}Nd/^{144}Nd$												
AGE	LC	LC	LC	Q	LC	LC	LC	LC	LC	LC	LC	PLI?
Reference	11	11	11	10	10	10	11	10	10	10	10	12

APPENDIX 7 (continued)

	Locality 64 Kyaukkadaung	Locality 64 Thuzi	Locality 64 Okpo Letpan	Locality 64 Natyindaung	Locality 66 Mt. Popa								
SiO_2	47.10	46.72	46.26	48.16	58.10	49.07	54.21	49.40	47.40	51.83	50.14	50.14	50.93
Al_2O_3	16.32	14.93	16.68	15.52	17.30	17.11	18.48	20.35	21.80	17.73	18.30	17.08	16.59
Fe_2O_3	3.80	5.12	7.77	6.25	7.20	5.10	6.06	10.05	10.05	3.82	2.99	4.03	4.69
FeO	5.00	3.98	2.41	4.10		3.29	1.16			4.16	5.43	5.36	3.93
MgO	11.65	8.70	5.17	8.22	3.15	6.43	4.02	4.78	5.68	5.70	6.02	6.22	6.09
CaO	11.45	12.87	13.43	10.53	7.40	10.72	8.08	12.10	12.25	10.44	10.66	10.87	10.79
Na_2O	1.95	2.12	3.06	2.84	2.22	2.79	2.98	2.50	1.96	2.87	2.59	2.56	2.52
K_2O	0.87	1.45	1.00	0.92	1.62	1.95	2.44	1.06	1.08	1.21	1.11	1.36	1.66
H_2O+	0.50	2.30	0.59	1.39		1.68	1.21			0.74			
H_2O-	0.40	0.55	0.19	0.77		0.81	0.44			0.04			
TiO_2	0.35	1.10	1.07	0.73		0.52	0.58			0.83	0.81	0.87	0.85
P_2O_5	0.18	0.27	0.31	0.30		0.45	0.33			0.21	0.17	0.22	0.21
MnO	0.25		0.16	0.16	0.25	0.18	0.13			0.14	0.14	0.14	0.13
BaO		0.08	0.03	0.05		0.08	0.14			0.08			
CO_2													
Ba											423	522	680
Sr											795	1005	865
Rb											12	18	22
Zr											76	90	85
Y											16	18	19
La											19	34	28
Ce											36	50	40
Nd													
Sm													
Nb											3	4	6
Li											6	6	6
B											0	1	1
U											2	0	0
Th											9	6	6
Mo											5	4	1
Pb											14	9	22
Zn											58	81	65
Cu											126	129	114
Ni											18	21	19
Cr											41	56	69
V													
Co											36	50	49
$^{87}Sr/^{86}Sr$											70435	70433	70445
$^{143}Nd/^{144}Nd$													
AGE	LC	LC	LC	LC	PLI	PLE	PLI	PLE	PLE	PLE	PLE	PLE	PLE
Reference	10	10	10	10	10	10	10	10	10	10	11	11	11

APPENDIX 7 (continued)

	Locality 66 Mt. Popa											Locality 67 Prome	Locality 68 Tharrawaddy	Locality 71 Kuantan	
SiO_2	53.26	50.23	52.86	56.32	58.84	60.20	58.74	58.35	59.97	64.44	64.40	48.00	49.70	47.45	39.35
Al_2O_3	16.10	17.04	17.09	17.20	16.56	17.24	18.77	17.38	17.40	15.16	17.85	18.70	18.04	13.68	12.23
Fe_2O_3	2.86	6.09	5.27	4.18	3.24	2.87	3.13	3.08	3.41	2.39	2.34	3.08	2.10	3.45	6.58
FeO	4.29	2.43	2.57	2.17	1.79	2.31	2.45	2.29	1.68	1.44	0.77	4.29	4.92	8.03	7.01
MgO	6.56	5.72	4.79	3.69	3.60	2.82	1.98	2.05	1.81	2.41	1.05	8.30	7.65	7.48	9.82
CaO	10.01	10.22	9.44	7.39	5.55	6.22	5.15	6.49	6.45	3.94	3.86	8.80	9.50	8.00	11.64
Na_2O	2.43	2.40	2.88	3.25	2.99	3.50	3.00	3.57	3.63	4.61	3.78	2.00	3.90	3.59	2.84
K_2O	1.50	1.95	2.17	2.66	2.95	2.54	2.40	3.08	2.85	2.79	3.64	0.72	1.35	1.46	1.40
H_2O+												3.70	1.80	3.11	4.85
H_2O-												1.70	0.10	1.05	1.11
TiO_2	0.72	0.93	0.75	0.64	0.54	0.10	0.22	0.56	0.59	0.12	0.01	0.30	0.45	1.85	2.14
P_2O_5	0.18	0.22	0.28	0.17	0.22	0.21	0.31	0.24	0.15	0.09	0.20	0.10	0.29	0.69	0.82
MnO	0.11	0.13	0.12	0.13	0.08	0.12	0.13	0.10	0.09	0.04	0.03	0.20	0.20	0.12	0.13
BaO														0.03	0.05
CO_2															
Ba	583	784	1069		650			640							
Sr	935	711	1050		562			508							
Rb	21	37	37		108			116							
Zr	82	103	120		165			180							
Y	13	24	21		21			24							
La	26	48	62		75			54							
Ce	41	49	71		70			67							
Nd															
Sm															
Nb	7	4	5		7			6							
Li	7	8	7		26			28							
B	0	0	0		25			22							
U	0	0	0		5			6							
Th	7	11	13		23			21							
Mo	2	0	2		2			2							
Pb	12	14	8		23			28							
Zn	59	64	61		70			66							
Cu	56	92	73		51			52							
Ni	30	27	17		16			12							
Cr	155	37	46		66			28							
V															
Co	42	50	36		20			18							
$^{87}Sr/^{86}Sr$	70431	70453	70445		70478			70501							
$^{143}Nd/^{144}Nd$															
AGE	PLE	PLE	PLE	PLI	PLI	PLI	PLI	PLI	PLI	PLI	PLI	LC	LC	PLE	PLE
Reference	11	11	11	11	11	11	11	11	11	11	11	10	10	13	13

APPENDIX 7 (continued)

	Locality 72 Segamat			Locality 73 Ban Chiang Khian			Locality 74 Lampang						Locality 76 Denchai			
SiO_2	47.74	4[illegible].75	49.28	48.90	49.20	50.00	49.30	46.90	47.20	48.00	46.30	44.40	47.50	49.10	45.90	45.70
Al_2O_3	8.98	1[illegible].30	15.79	15.80	16.40	16.00	16.00	14.90	15.90	15.40	16.00	13.70	14.50	14.70	15.70	14.60
Fe_2O_3	5.56	[illegible].38	6.37	2.50	3.70	2.00	3.70	3.40	1.80	2.20	2.10	2.40	3.10	2.80	2.00	4.30
FeO	5.48	[illegible].20	4.37	9.50	7.30	9.30	5.30	6.10	5.90	6.10	5.50	4.90	6.50	6.90	7.50	5.70
MgO	10.59	[illegible].95	8.56	8.40	5.30	5.80	7.10	10.20	6.60	8.20	6.00	7.30	7.60	7.30	8.40	9.40
CaO	11.96	7.40	4.70	8.10	8.70	9.40	7.30	7.40	9.30	8.10	9.60	11.70	7.40	7.40	8.20	7.80
Na_2O	1.31	[illegible].29	2.68	2.70	3.00	2.90	3.80	3.10	4.10	3.30	3.40	4.50	3.60	3.50	3.40	2.90
K_2O	3.78	7.44	6.12	0.60	1.20	0.80	3.00	3.00	3.00	3.10	3.40	1.40	1.60	1.60	2.50	2.90
H_2O+	2.38	[illegible].53	1.64	1.30	1.00	0.30			2.00	2.20	2.40	1.00	2.90	2.30	2.10	1.60
H_2O-	0.76	[illegible].38		0.50	1.40	0.60			0.40	0.40	0.70	1.50	1.40	1.20	1.00	2.00
TiO_2	0.97	[illegible].82	0.95	1.50	1.91	1.43	2.24	2.14	2.30	2.30	2.30	2.10	1.77	1.85	2.10	2.26
P_2O_5	0.04			0.30	0.34	0.22	0.75	0.69	0.70	0.60	0.70	0.60	0.30	0.36	0.55	0.61
MnO	0.18			0.14	0.20	0.21	0.16	0.16	0.15	0.15	0.17	0.19	0.15	0.15	0.14	0.18
BaO																
CO_2				0.20	0.70	0.80	1.10	2.10	0.90	0.20	1.20	3.80	0.40	0.30	0.20	0.30
Ba													254	276	376	399
Sr													442	971	659	692
Rb													56	28	30	40
Zr													138	163	197	208
Y																
La													14	17	25	24
Ce													27	31	43	45
Nd																
Sm																
Nb													24	26	40	44
Li																
B																
U																
Th																
Mo																
Pb																
Zn																
Cu																
Ni																
Cr																
V																
Co																
$^{87}Sr/^{86}Sr$																
$^{143}Nd/^{144}Nd$																
AGE	PAL	PA[illegible]	PAL	PLE	PLE	PLE	PLE	PLE	PLE	PLE	PLE	PLE	MI?	MI?	MI?	MI?
Reference	13	13	14	15	16	16	16	16	15	15	15	15	17	17	17	17

APPENDIX 7 (continued)

	Locality 76 Denchai												Locality 76 Mae Lama	Locality 77 Bo Phloi	
SiO_2	46.40	47.80	47.60	50.90	49.30	49.50	49.80	50.90	50.90	51.20	50.30	49.10	50.00	45.40	45.90
Al_2O_3	15.20	14.70	15.70	14.80	14.80	15.30	14.70	16.30	16.50	14.40	14.60	15.70	16.50	13.20	13.30
Fe_2O_3	2.50	2.90	1.70	1.60	2.20	2.10	2.20	2.90	3.80	3.30	2.70	2.30		3.40	2.80
FeO	6.80	6.40	8.20	8.50	7.10	7.20	7.20	5.30	4.50	4.50	5.70	7.20	9.90	6.30	6.60
MgO	7.10	8.70	8.40	7.20	8.30	8.60	8.50	6.80	6.40	6.00	6.80	7.70	6.60	9.00	9.20
CaO	8.80	7.70	8.60	7.60	7.60	7.30	8.40	6.90	7.00	7.40	7.30	7.80	8.40	9.30	8.80
Na_2O	2.80	2.50	2.30	2.80	2.60	1.90	2.50	3.10	3.20	3.20	3.90	3.60	3.00	5.20	5.00
K_2O	3.20	1.80	1.90	2.00	2.40	2.30	1.90	2.30	2.40	1.90	1.60	1.40	1.60	2.50	2.40
H_2O+	1.80	2.60	2.50	0.90	1.80	1.50	2.10	2.20	1.60	0.90	1.70	2.10		2.00	2.40
H_2O-	1.50	1.90	0.40	0.90	1.20	1.40	1.00	1.40	2.10	2.30	1.20	1.20		0.80	0.30
TiO_2	2.16	1.86	1.80	1.91	1.84	1.87	1.88	1.70	1.70	1.67	1.67	1.97	1.40	2.00	1.70
P_2O_5	0.69	0.44	0.50	0.39	0.42	0.46	0.41	0.50	0.50	0.47	0.51	0.50	0.45	0.80	0.80
MnO	0.18	0.16	0.15	0.17	0.15	0.18	0.16	0.10	0.12	0.11	0.15	0.16	0.14	0.18	0.19
BaO															
CO_2	0.90	0.40	0.20	0.20	0.20	0.20	0.20	0.10		0.50	0.70	0.30	2.80	0.10	0.10
Ba	433	306	287	267	291	302	276		362	408	466	381			
Sr	778	811	555	569	692	573	535		549	596	642	588			
Rb	33	23	32	34	44	34	27		46	13	41	24			
Zr	227	189	169	189	177	173	164		179	182	178	190			
Y															
La	27	18	17	17	19	17	17		19	20	25	23			
Ce	50	34	32	27	33	34	34		40	39	42	44			
Nd															
Sm															
Nb	42	30	37	36	30	30	31		35	30	25	36			
Li															
B															
U															
Th															
Mo															
Pb															
Zn															
Cu															
Ni															
Cr															
V															
Co															
$^{87}Sr/^{86}Sr$															
$^{143}Nd/^{144}Nd$															
AGE	MI?	MI?	MI?	MI?	MI?	MI?	MI?	MI?	MI?	MI?	MI?	MI?	LC	LC	LC
Reference	17	17	17	17	17	17	17	17	17	17	17	17	16	15	15

APPENDIX 7 (continued)

	Locality 78 Kok Sam Rong				Locality 79 Nakhon Ratchasima		Locality 80 Pradonchai			Locality 81 Buriram	Locality 82 Surin	Locality 83 Kantharalak			
SiO_2	47.10	48.00	52.70	53.10	48.10	49.10	51.50	52.40	51.50	48.90	52.20	49.52	50.10	50.60	48.26
Al_2O_3	16.30	16.30	15.90	16.20	15.20	14.90	14.90	14.40	14.10	13.50	14.50	17.65	15.10	15.70	17.05
Fe_2O_3	2.20	4.30	5.50	6.40	6.00	7.80	5.40	2.90	4.10	4.10	3.80	7.90	2.30	6.60	11.23
FeO	7.40	7.90	3.00	1.20	5.20	4.30	3.40	6.40	5.10	6.50	5.05		6.90	4.10	
MgO	7.90	5.50	6.30	4.00	5.10	5.30	4.20	5.50	5.10	6.80	5.70	4.54	5.70	6.00	5.49
CaO	10.10	7.20	8.70	7.80	9.10	9.50	6.00	7.20	6.80	7.70	6.50	7.66	7.60	7.10	9.40
Na_2O	3.10	4.00	3.90	4.50	3.30	4.00	4.80	4.30	4.00	3.70	4.90	4.84	5.20	3.50	4.17
K_2O	0.70	0.90	1.20	2.30	1.40	1.20	2.30	1.50	1.40	1.80	1.70	3.81	2.60	1.50	1.43
H_2O+	3.30	2.60	0.70	1.30	1.10	0.90	1.60	1.40	1.30	1.10	1.50		2.00	1.40	
H_2O-	0.80	1.40	0.60	2.70	1.70	1.00	1.50	1.80	2.40	1.70	1.60		0.30	1.20	
TiO_2	1.40	2.10	1.30	1.60	2.20	2.10	3.00	2.80	2.80	2.90	2.40	1.82	2.20	1.65	2.18
P_2O_5	0.20	0.30	0.30	0.50	0.40	0.50	1.10	0.70	0.80	0.70	0.60	0.50	0.70	0.40	0.19
MnO	0.16	0.18	0.14	0.13	0.17	0.15	0.10	0.14	0.11	0.14	0.11	0.22	0.13	0.07	0.25
BaO															
CO_2	0.10	0.10	0.10	0.10	0.10	0.10		0.10	0.30	0.10	0.20		0.20		
Ba															
Sr															
Rb															
Zr															
Y															
La															
Ce															
Nd															
Sm															
Nb															
Li															
B															
U															
Th															
Mo															
Pb															
Zn															
Cu															
Ni															
Cr															
V															
Co															
$^{87}Sr/^{86}Sr$															
$^{143}Nd/^{144}Nd$															
AGE	LC	LC	LC	LC	LC	LC	LC	LC	LC	LC	LC	LC	LC	LC	LC
Reference	15	15	15	15	15	15	15	15	15	15	15	15	15	15	15

APPENDIX 7 (continued)

	Locality 84 Tha Mai		Locality 84 Chanthaburi						Locality 84 Pong Nam Rong	Locality 85 Ben Huai Sai		Locality 86 Bokeo			
SiO_2	40.60	40.20	39.90	42.00	40.70	40.90	41.30	40.00	42.60	45.44	46.70	43.40	42.65	42.90	45.55
Al_2O_3	10.80	11.60	11.80	11.70	12.50	12.20	12.60	12.20	12.20	16.14	15.80	13.35	12.25	11.65	10.90
Fe_2O_3	5.20	4.10	6.60	11.90	3.90	3.80	3.70	5.90	2.70	3.89		5.15	7.45	5.30	6.75
FeO	8.50	9.60	7.40	2.40	8.50	9.40	9.30	8.40	8.20	7.22	11.50	8.25	6.50	8.25	7.70
MgO	10.40	9.60	9.10	8.70	9.50	8.30	8.30	8.20	8.20	9.02	6.40	10.90	11.70	12.30	9.40
CaO	10.10	10.10	10.00	10.00	10.20	10.10	9.90	10.40	12.30	8.98	8.50	9.75	8.90	9.70	9.60
Na_2O	4.30	4.90	5.10	3.50	3.30	4.30	4.50	4.60	2.70	3.02	3.40	2.75	2.00	3.00	2.70
K_2O	2.40	2.20	1.20	0.90	1.50	2.00	1.90	1.80	2.30	1.63	3.40	1.85	1.35	2.10	1.60
H_2O+	2.40	2.20	2.60	2.40	4.10				2.90	1.15		1.65	3.45	2.00	0.60
H_2O-	0.70	0.60	1.00	1.90	0.80				0.20	0.72		0.45	1.25	0.50	0.40
TiO_2	3.30	3.10	3.17	2.55	3.02	3.21	3.16	3.40	2.80	2.22	2.40	2.05	1.95	1.75	2.65
P_2O_5	1.10	1.10	1.10	1.01	0.65	0.94	0.91	0.93	0.80	0.32	0.98	0.90	1.10	1.40	1.30
MnO	0.20	[illegible].16	0.20	0.20	0.19	0.20	0.20	0.20	0.19	0.23	0.18	0.15	0.15	0.10	0.20
BaO															
CO_2	0.20	[illegible].20	0.50	0.10	0.40	2.50	2.20	2.30	2.50		2.20				
Ba															
Sr															
Rb															
Zr															
Y															
La															
Ce															
Nd															
Sm															
Nb															
Li															
B															
U															
Th															
Mo															
Pb															
Zn															
Cu															
Ni															
Cr															
V															
Co															
$^{87}Sr/^{86}Sr$															
$^{143}Nd/^{144}Nd$															
AGE	PLE	PLE	PLE	PLE	PLE	PLE	PLE	PLE	PLI?	LC	PLE	Q	Q	Q	Q
Reference	15	15	16	16	16	16	16	16	15	15	16	18	18	18	18

APPENDIX 7 (continued)

	Locality 86 Bokeo				Locality 91 Pailin								Locality 99 Pleiku		
SiO_2	46.20	45.60	49.10	46.90	47.00	43.50	43.70	41.50	43.10	41.50	45.80	45.50	52.75	52.66	49.40
Al_2O_3	11.30	10.10	14.70	9.70	10.35	13.91	13.20	12.80	13.10	12.70	14.20	14.20	12.35	13.84	13.80
Fe_2O_3	6.50	6.95	4.50	4.85	9.40	12.12	7.60	5.70	6.50	9.80	3.80	4.10	7.45	10.02	6.20
FeO	6.10	6.10	6.40	7.40	4.05		5.20	7.20	6.80	4.20	6.80	6.80			5.10
MgO	12.20	11.55	5.30	10.50	9.90	8.48	7.40	9.30	8.40	7.60	6.90	7.10	8.80	6.52	7.70
CaO	10.35	9.05	6.40	8.85	9.20	10.78	10.10	10.30	9.80	10.10	8.50	8.60	9.45	8.94	8.30
Na_2O	2.90	2.75	6.60	2.70	2.40	3.30	3.30	3.60	4.40	3.60	4.80	5.00	2.50	2.60	3.40
K_2O	2.70	1.90	3.10	1.40	1.40	2.20	3.30	1.40	0.90	1.20	2.70	2.60	0.50	0.80	1.90
H_2O+		1.60		3.50	1.80	0.78	3.10	3.50			1.90	1.60	3.40	1.06	0.70
H_2O-	0.3[illegible]	1.70	0.05	0.95	1.00	0.43	1.90	0.70			0.80	0.80	0.40	1.13	0.70
TiO_2	2.7[illegible]	1.90	2.00	2.40	2.55	2.82	2.64	3.15	3.03	3.35	2.52	2.47	1.60	1.98	2.12
P_2O_5	1.0[illegible]	1.00	1.30	0.70	1.00	0.84	1.01	0.85	0.85	1.20	0.77	0.80	0.70	0.28	0.74
MnO	0.2[illegible]	0.15	0.10	0.10	0.20	0.15	0.22	0.19	0.19	0.19	0.19	0.19	0.15	0.20	0.17
BaO															
CO_2							0.30	0.60	2.20	4.50	0.50	0.40			0.10
Ba															
Sr															
Rb															
Zr															
Y															
La															
Ce															
Nd															
Sm															
Nb															
Li															
B															
U															
Th															
Mo															
Pb															
Zn															
Cu															
Ni															
Cr															
V															
Co															
$^{87}Sr/^{86}Sr$															
$^{143}Nd/^{144}Nd$															
AGE	Q	Q	Q	Q	Q	Q	PLI	PLE	PLE	PLE	PLE	PLE	PLI?	PLI?	PLI
Reference	18	18	18	18	18	18	16	16	16	16	16	16	18	18	16

APPENDIX 7 (continued)

	Locality 101 Darlac	Locality 103 Bu Kong Rong			Locality 104 Lien Khang		Locality 107 Xuan Loc	Locality 117 Khubsugol			Locality 122 Narim-Gichegene				Locality 122 Khorog
SiO_2	47.40	53.40	50.90	49.50	50.00	50.34	46.40	48.34	47.40	49.72	45.50	47.00	47.00	50.60	46.00
Al_2O_3	14.10	15.80	14.40	14.40	12.05	12.85	12.00	15.38	15.50	15.38	16.60	14.60	16.60	14.20	14.70
Fe_2O_3	2.00	4.70	3.40	1.60	11.90	11.50	2.00	3.48	3.83	5.03	3.48	3.22	3.23	1.50	1.49
FeO	9.20	4.00	7.60	9.60			9.40	7.46	7.31	6.09	7.59	7.28	7.00	8.83	9.29
MgO	8.20	3.50	7.40	7.80	8.30	5.71	11.20	7.30	7.92	6.65	6.30	6.25	5.90	6.00	7.90
CaO	8.30	5.00	7.60	8.70	10.00	8.98	7.90	7.71	7.88	7.69	6.80	7.50	7.20	6.80	7.60
Na_2O	3.20	4.80	3.20	2.90	2.80	2.80	2.70	3.26	6.51	3.30	5.00	4.40	4.26	4.90	4.50
K_2O	1.50	3.20	1.00	1.20	0.90	1.36	2.10	3.14	2.20	2.37	5.00	4.10	4.10	4.80	4.10
H_2O+	2.40	1.50	1.30	0.90	0.80	1.83	0.70	0.88	0.73	0.44					
H_2O-	0.90	1.80	1.70	0.60	0.20	2.11	0.50		0.33	0.05					
TiO_2	1.93	2.11	1.63	2.16	2.00	2.08	2.49	2.07	2.48	2.29	2.47	2.32	2.52	2.24	2.38
P_2O_5	0.44	0.62	0.23	0.38	0.70	0.42	0.61		0.80	0.65					
MnO	0.17	0.11	0.15	0.17	0.15	0.25	0.18		0.16	0.15	0.19	0.18	0.18	0.17	0.24
BaO															
CO_2	0.10	0.10	0.10	0.30			0.30								
Ba															
Sr															
Rb															
Zr															
Y															
La															
Ce															
Nd															
Sm															
Nb															
Li															
B															
U															
Th															
Mo															
Pb															
Zn															
Cu															
Ni															
Cr															
V															
Co															
$^{87}Sr/^{86}Sr$															
$^{143}Nd/^{144}Nd$															
AGE	PLI	MI	MI	MI	Q?	Q?	PLI	LC	LC	LC	H	H	H	H	H
Reference	16	16	16	16	18	18	16	19	20	20	21	21	21	21	21

APPENDIX 7 (continued)

	Locality 122 Khorog			Locality 122 Chulutu 40-90 m terrace			Locality 122 Chulutu 150-200 m terrace									
SiO_2	47.50	49.00	50.25	47.00	47.70	49.50	45.30	45.50	46.70	46.80	47.70	48.00	51.00	53.00	50.50	
Al_2O_3	14.30	16.60	14.90	14.20	13.35	14.75	13.50	14.80	12.90	13.10	13.60	14.40	15.10	16.45	16.00	
Fe_2O_3	0.91	2.75	1.32	2.63	2.20	1.51	2.63	1.64	0.87	2.01	2.99	0.79	1.88	6.36	3.82	
FeO	10.22	6.89	8.18	7.72	8.33	8.33	8.76	8.25	9.31	8.91	7.94	8.93	7.50	4.45	3.10	
MgO	7.50	6.60	7.40	8.40	8.40	8.80	10.70	9.20	8.90	9.80	8.22	7.10	7.20	4.30	4.40	
CaO	8.40	7.00	6.00	8.50	8.10	6.82	7.75	7.70	8.93	7.95	8.05	8.80	7.35	9.85	6.95	
Na_2O	4.20	4.00	4.50	3.98	3.98	4.46	4.36	3.66	4.26	4.36	3.45	4.16	4.04	3.45	3.82	
K_2O	2.90	3.10	4.04	3.06	3.25	3.74	2.98	3.55	3.18	2.98	2.64	3.27	2.48	0.16	2.16	
H_2O+																
H_2O-																
TiO_2	2.12	2.20	2.28	2.70	2.52	2.25	2.45	2.54	2.50	2.28	2.30	2.56	2.47	1.33	1.02	
P_2O_5																
MnO	0.20	0.15	0.10	0.20	0.19	0.20	0.23	0.22	0.18	0.22	0.19	0.19	0.19	0.19	0.12	
BaO																
CO_2																
Ba																
Sr																
Rb																
Zr																
Y																
La																
Ce																
Nd																
Sm																
Nb																
Li																
B																
U																
Th																
Mo																
Pb																
Zn																
Cu																
Ni																
Cr																
V																
Co																
$^{87}Sr/^{86}Sr$																
$^{143}Nd/^{144}Nd$																
AGE	H	H	H	PLI	PLI	PLI	PLE	PLE	PLE	PLE	PLE	PLE	PLE	PLE	PLE	PLE
Reference	21	21	21	21	21	21	21	21	21	21	21	21	21	21	21	21

APPENDIX 7 (continued)

	Locality 122 Chulutu lava feeders											Locality 123 Khangay				Locality 124 Gobi Altai
SiO_2	49.50	53.0[illegible]	45.00	48.50	45.00	46.10	46.50	47.10	48.00	48.00	49.50	48.40	49.63	51.25	45.60	48.39
Al_2O_3	13.30	14.7[illegible]	15.60	13.40	14.70	16.10	13.30	15.70	13.60	13.70	14.40	14.62	14.79	17.20	17.66	14.78
Fe_2O_3	2.89	3.5[illegible]	3.64	2.85	3.39	4.02	4.05	4.74	4.00	3.70	6.00	2.01	2.33	2.75	3.84	6.08
FeO	7.27	5.1[illegible]	7.35	6.98	7.00	6.06	6.98	5.19	6.06	6.31	4.23	8.81	8.96	7.25	7.12	4.64
MgO	9.00	5.0[illegible]	8.90	8.50	8.90	6.50	8.90	8.25	7.80	9.00	7.50	8.35	7.16	5.74	4.44	6.72
CaO	8.15	7.5[illegible]	8.90	8.10	8.30	7.45	8.10	7.13	7.02	8.15	7.15	5.49	5.49	5.54	6.93	7.36
Na_2O	3.10	4.4[illegible]	3.00	3.96	4.22	5.05	4.22	2.77	5.05	4.68	2.94	3.97	3.98	3.73	5.85	4.17
K_2O	2.09	2.4[illegible]	2.58	2.24	2.10	2.98	2.12	1.93	2.10	1.99	2.93	4.48	3.99	4.12	4.56	2.42
H_2O+																1.53
H_2O-																
TiO_2	2.22	2.6[illegible]	2.62	2.32	2.27	2.28	2.51	2.23	2.26	2.24	2.24	2.25	2.20	2.07	2.65	2.39
P_2O_5												0.80	0.75			
MnO	0.20	0.1[illegible]	0.22	0.20	0.21	0.20	0.20	0.17	0.20	0.20	0.17			0.13	0.15	
BaO																
CO_2																
Ba																
Sr																
Rb																
Zr																
Y																
La																
Ce																
Nd																
Sm																
Nb																
Li																
B																
U																
Th																
Mo																
Pb																
Zn																
Cu																
Ni																
Cr																
V																
Co																
$^{87}Sr/^{86}Sr$																
$^{143}Nd/^{144}Nd$																
AGE	PLI	PLI	PLI	PLI	PLI	PLI	PLI	PLI	PLI	PLI	PLI					
Reference	21	21	21	21	21	21	21	21	21	21	21	22	22	22	22	19

APPENDIX 7 (continued)

	Locality 125 Dariganga	Locality 127 Alpha	Locality 127 Gamma		Locality 127 Beta	Locality 127 Aluchin lavas						Locality 128 Anyuy		
SiO_2	45.28	50.27	47.69	46.54	53.98	51.21	52.98	49.54	50.18	49.40	49.77	52.27	52.33	53.60
Al_2O_3	11.84	16.98	15.75	16.46	17.19	16.01	16.20	15.59	18.62	15.34	15.96	15.56	15.53	15.55
Fe_2O_3	6.13	10.69	10.39	6.99	1.92	8.75	3.05	3.61		1.81	1.56	1.62	1.88	5.26
FeO	2.21	0.66	0.95	4.42	5.42	1.90	6.75	9.41	7.21	9.41	9.05	8.05	7.26	4.62
MgO	10.03	5.51	6.29	6.39	5.51	7.23	7.00	7.61	7.69	7.44	7.25	6.48	6.56	5.91
CaO	9.10	4.91	6.48	5.90	5.83	6.99	7.25	6.83	8.59	7.20	7.17	8.03	7.20	7.20
Na_2O	3.35	3.72	4.00	3.48	5.06	3.21	4.08	2.64	3.36	3.75	3.92	3.60	3.66	3.50
K_2O	1.72	2.22	2.29	2.31	1.26	2.01	1.51	1.53	1.63	1.60	1.87	1.87	2.06	1.46
H_2O+	1.01	0.50	0.98	1.61	0.23	0.17	0.28	0.21	0.17	0.26	0.38	0.13	0.29	0.60
H_2O-												0.11	0.16	0.26
TiO_2	2.79	1.78	2.07	2.07	0.63	1.60	0.48	1.28	1.80	2.05	1.84	1.67	1.58	1.49
P_2O_5		0.54	0.68	0.56	0.14	0.39	0.33	0.36	0.26	0.40	0.34	0.59	0.45	0.41
MnO	0.18	0.16	0.18	0.19	0.18	0.15	0.14	0.10	0.14	0.11	0.15	0.12	0.14	0.13
BaO														
CO2														
Ba														
Sr														
Rb														
Zr														
Y														
La														
Ce														
Nd														
Sm														
Nb														
Li														
B														
U														
Th														
Mo														
Pb														
Zn														
Cu														
Ni														
Cr														
V														
Co														
$^{87}Sr/^{86}Sr$														
$^{143}Nd/^{144}Nd$														
AGE		H	H	H	H	H	H	H	H	H	H	H	H	H
Reference	19	23	23	23	23	23	23	23	23	23	23	24	24	24

APPENDIX 7 (continued)

	Locality 152 S. Sikhote-Alin			Locality 153 W. Sakukan		Locality 153 Dvoinoi	Locality 153 Yakutsk						Locality 153 Udokan Plateau		
SiO_2	65.44	66.26	74.41	42.80	41.28	42.56	49.60	47.30	49.96	50.16	50.30	50.10	47.66	50.32	49.24
Al_2O_3	15.17	15.43	13.49	14.85	13.13	14.04	16.77	15.92	16.67	16.36	16.20	16.11	15.64	17.13	16.93
Fe_2O_3	5.04	4.22	1.92	3.27	9.08	4.48	7.13	7.54	4.99	6.64	7.50	7.47	4.36	3.53	4.57
FeO	2.17	1.10	0.60	10.55	6.03	8.84	2.08	1.07	4.45	3.30	2.37	2.66	7.73	7.73	7.39
MgO	4.05	2.74	0.62	8.70	9.06	10.60	4.28	4.44	4.09	4.09	4.01	4.24	6.53	3.35	4.49
CaO	0.05	1.23	0.28	10.02	10.27	9.90	5.08	5.18	5.03	5.25	5.25	5.25	8.56	5.65	7.02
Na_2O	3.74	3.71	0.48	4.15	3.55	4.00	6.55	5.85	7.50	6.65	5.55	6.55	3.62	5.05	3.55
K_2O	4.39	3.72	5.30	1.92	1.26	1.80	3.20	3.50	3.40	3.00	3.00	3.40	1.46	2.86	1.72
H_2O^+													1.31	0.80	1.52
H_2O^-															
TiO_2	0.57	0.60	0.17	3.15	3.00	2.87	1.50	1.40	1.75	1.30	1.90	1.68	2.38	2.32	2.02
P_2O_5	0.02	0.01	0.10	0.69	0.71	0.69	0.46	0.46	0.46	0.42	0.42	0.44	0.50	1.05	0.92
MnO	0.20	0.14	0.05	0.16	0.16	0.14	0.16	0.16	0.13	0.14	0.12	0.14	0.16	0.16	0.20
BaO															
CO_2															
Ba															
Sr															
Rb															
Zr															
Y															
La															
Ce															
Nd															
Sm															
Nb															
Li															
B															
U															
Th															
Mo															
Pb															
Zn															
Cu															
Ni															
Cr															
V															
Co															
$^{87}Sr/^{86}Sr$															
$^{143}Nd/^{144}Nd$															
AGE	PAL	PAL	PAL	Q	Q	Q	Q	Q	Q	Q	Q	Q	NE	NE	NE
Reference	25	25	25	26	26	26	26	26	26	26	26	26	27	27	27

APPENDIX 7 (continued)

	Locality 153 Udokan Plateau									Locality 153 Udokan Ridge						
SiO_2	55.15	55.19	59.64	44.65	48.29	44.55	49.21	60.53	60.10	44.48	46.30	48.56	49.92	50.05	55.35	58.99
Al_2O_3	18.51	18.69	17.10	14.44	15.65	13.13	17.77	18.21	18.07	13.95	15.55	15.89	15.38	17.18	17.88	17.85
Fe_2O_3	4.11	2.95	4.58	3.83	3.17	4.51	1.53	3.64	2.74	8.23	4.44	5.63	5.39	4.30	2.64	2.63
FeO	3.64	2.98	2.36	9.34	8.82	9.29	4.48	1.96	2.80	8.22	8.67	6.46	6.53	7.27	5.20	4.78
MgO	1.84	0.66	1.46	9.64	6.77	10.18	0.76	0.47	0.57	9.51	5.88	6.03	7.50	3.68	2.17	0.62
CaO	4.83	2.12	2.89	9.54	8.89	8.30	2.03	1.79	1.88	8.95	7.81	8.30	9.16	6.04	3.58	1.85
Na_2O	5.64	6.77	4.50	3.54	3.50	1.54	6.82	6.71	6.75	3.70	4.12	3.98	3.17	4.83	5.62	6.42
K_2O	3.85	5.38	3.80	1.52	1.45	2.63	5.63	5.64	5.64	1.79	2.09	1.70	0.96	2.62	4.15	5.12
H_2O^+	0.02	0.50	1.40	0.67	0.92	0.30	1.14	0.64	0.80	0.61	0.88	0.55	0.29	0.47	0.41	0.58
H_2O^-										0.13	0.26	0.25	0.77	0.36	0.22	0.33
TiO_2	1.45	0.63	0.77	2.49	1.97	2.13	0.65	0.62	0.63	2.80	3.30	2.21	1.64	2.34	1.20	0.40
P_2O_5	0.50	0.17	0.89	0.52	0.37	0.14	0.10	0.10	0.10	0.33	0.43	0.48	0.30	0.83	1.07	0.28
MnO	0.16	0.16	0.17	0.18	0.17	0.20	0.15	0.14	0.15	0.17	0.18	0.16	0.18	0.09	0.18	0.24
BaO																
CO_2																
Ba																
Sr																
Rb																
Zr																
Y																
La																
Ce																
Nd																
Sm																
Nb																
Li																
B																
U																
Th																
Mo																
Pb																
Zn																
Cu																
Ni																
Cr																
V																
Co																
$^{87}Sr/^{86}Sr$																
$^{143}Nd/^{144}Nd$																
AGE	NE	NE	NE	PLE	PLE	PLE	PLE	H	H							
Reference	27	27	27	27	27	27	27	27	27	20	20	20	20	20	20	20

APPENDIX 7 (continued)

	Locality 157 Vitim Plateau															
SiO_2	49.38	50.22	50.78	45.33	48.93	47.34	50.36	50.17	49.32	50.07	50.21	48.42	49.16	47.92	48.20	48.44
Al_2O_3	15.05	15.51	14.98	14.33	15.54	13.63	15.26	15.35	15.25	15.24	15.21	15.41	15.61	15.26	14.73	14.59
Fe_2O_3	4.32	2.28	4.07	3.67	3.43	2.50	2.28	2.26	3.37	3.07	4.68	3.38	4.37	2.88	3.70	5.70
FeO	8.28	9.18	7.88	9.25	8.31	9.08	9.23	9.37	8.36	9.15	7.96	8.04	7.20	8.34	7.95	6.94
MgO	6.93	7.52	6.95	8.15	7.34	8.66	7.61	7.31	7.53	7.04	7.77	8.09	8.58	9.76	9.06	8.06
CaO	9.10	9.27	8.98	7.92	8.09	8.34	9.12	9.32	9.14	8.76	9.36	9.12	9.04	8.84	8.56	9.10
Na_2O	2.72	2.92	2.60	4.52	3.74	3.61	2.90	2.96	2.94	2.74	3.17	2.75	3.01	2.80	3.09	3.14
K_2O	1.02	1.06	1.12	2.38	1.37	2.09	1.24	1.10	1.32	1.20	1.46	1.72	1.83	1.53	1.65	1.69
H_2O+	0.64	0.10	0.20	0.83	0.80		0.08	0.17	0.40	0.62	0.04	0.17	0.15	0.05	0.14	0.16
H_2O-				0.22	0.35											
TiO_2	2.08	1.92	2.08	2.49	2.12	2.38	1.92	1.92	2.08	2.08	1.20	1.30	1.30	1.60	1.70	2.00
P_2O_5	0.34	0.26	0.31	0.55	0.34	0.58	0.25	0.25	0.29	0.25	0.35	0.45	0.48	0.41	0.45	0.43
MnO	0.15	0.15	0.16	0.15	0.14	0.16	0.15	0.15	0.16	0.15	0.15	0.14	0.12	0.12	0.13	0.11
BaO																
CO_2																
Ba																
Sr																
Rb																
Zr																
Y																
La																
Ce																
Nd																
Sm																
Nb																
Li																
B																
U																
Th																
Mo																
Pb																
Zn																
Cu																
Ni																
Cr																
V																
Co																
$^{87}Sr/^{86}Sr$																
$^{143}Nd/^{144}Nd$																
AGE	Q	Q	Q				Q	Q	Q	Q	Q	Q	Q	Q	Q	Q
Reference	28	28	28	20	20	29	28	28	28	28	28	28	28	28	28	28

APPENDIX 7 (continued)

	Locality 158 Dzhida				Locality 159 Khamar-Daban			Locality 160 Tunka			Locality 161 Kitoy River		Locality 163 Oka River		Locality 165 NE Tuva (N. field)
SiO_2	47.51	45.71	47.45	46.93	47.45	46.67	47.42	46.93	46.24	47.25	44.54	46.55	47.71	48.08	45.92
Al_2O_3	16.42	15.91	14.50	14.12	14.40	14.72	14.82	16.28	15.97	14.39	14.52	14.67	14.67	14.55	14.56
Fe_2O_3	5.32	5.13	3.60	2.11	3.69	2.34	5.43	4.76	4.46	3.02	2.86	4.39	3.72	4.90	6.06
FeO	6.40	7.33	9.84	9.23	9.84	9.89	6.16	7.20	7.23	8.42	9.23	7.58	7.49	6.98	4.31
MgO	6.06	6.61	8.14	8.39	8.14	8.30	7.47	7.84	8.24	8.41	8.81	8.60	7.64	7.36	6.52
CaO	7.50	7.09	8.39	8.46	8.39	8.37	7.65	8.34	8.59	8.57	8.75	8.79	8.33	7.47	6.73
Na_2O	4.13	5.07	2.38	3.50	2.38	3.94	3.06	3.51	2.01	3.06	2.80	1.76	3.36	3.18	2.46
K_2O	2.32	2.37	1.47	2.10	1.47	1.97	1.42	1.69	1.22	1.81	1.35	0.88	1.96	1.85	1.37
H_2O+	1.18	0.87	1.15			0.86	3.43	0.65	2.15		3.72	2.07	0.65	1.51	4.66
H_2O-						0.20	1.21	0.04	0.45			1.29	0.32	0.19	4.05
TiO_2	2.26	2.53	1.96	2.51	1.96	2.84	2.83	2.11	2.28	2.40	2.42	2.15	2.56	2.56	2.23
P_2O_5		0.72	0.70	0.72	0.70	0.36	0.44	0.36	0.51	0.65	0.22	0.54	0.65	0.55	0.74
MnO	0.15	0.15	0.17	0.16	0.17	0.15	0.16	0.26	0.28	0.16	0.14	0.13	0.13	0.14	0.13
BaO															
CO_2															
Ba															300
Sr															700
Rb															<10
Zr															60
Y															
La															
Ce															
Nd															
Sm															
Nb															14
Li															9.8
B															
U															
Th															
Mo															1.5
Pb															7
Zn															130
Cu															34
Ni															82
Cr															150
V															150
Co															42
Sn															3
Ta															0.8
Ga															18
$^{87}Sr/^{86}Sr$															
$^{143}Nd/^{144}Nd$															
AGE		LC	LC												
Reference	19	20	20	29	29	20	20	20	20	29	20	20	20	20	30

APPENDIX 7 (continued)

	Locality 165 NE Tuva, North field																
SiO_2	47.66	48.31	48.54	48.58	48.59	48.65	48.69	48.78	48.83	49.06	49.42	49.50	49.54	49.58	49.74	48.94	49.79
Al_2O_3	16.52	14.23	14.80	15.31	13.53	14.40	13.75	14.34	12.90	15.30	15.09	15.00	15.06	14.69	15.54	15.08	14.52
Fe_2O_3	3.28	6.16	3.12	3.36	5.97	3.94	5.83	4.86	8.59	4.38	3.20	3.90	2.49	4.78	5.45	3.12	2.96
FeO	6.54	4.67	6.68	6.46	4.13	6.46	5.56	4.74	3.05	5.10	6.97	5.56	7.18	4.81	3.59	6.46	5.75
MgO	7.19	7.15	8.58	7.54	7.56	7.28	6.90	9.43	6.80	7.86	7.37	7.05	7.09	7.90	6.72	8.78	6.56
CaO	7.80	8.50	7.29	7.24	7.50	8.10	8.13	6.68	7.38	6.85	6.96	7.58	7.02	7.13	7.20	6.68	7.26
Na_2O	3.40	3.56	3.04	3.58	3.46	3.56	3.56	2.62	3.41	2.88	3.46	3.46	3.58	3.46	3.56	3.40	3.36
K_2O	2.16	2.27	1.80	2.16	2.27	2.40	2.21	1.84	2.40	1.80	2.00	2.40	2.70	1.26	2.27	1.92	2.27
H_2O+	2.06	1.66	1.28	1.52	2.32	1.62	1.77	1.80	1.87	1.78	1.24	1.82	0.90	1.54	1.46	1.24	3.16
H_2O-	0.24	0.68	0.62	0.44	1.94	0.78	0.68	0.84	1.73	0.66	0.34	0.92	0.38	0.86	1.61	0.14	1.64
TiO_2	2.30	2.03	2.19	2.19	2.34	2.36	2.11	2.11	2.19	2.19	2.14	2.30	2.19	2.16	2.27	2.16	2.23
P_2O_5	0.75	0.74	0.74	0.74	0.70	0.70	0.74	0.75	0.74	0.71	0.74	0.70	0.73	0.73	0.66	0.75	0.63
MnO	0.10	0.17	0.10	0.10	0.14	0.19	0.18	0.13	0.14	0.10	0.11	0.15	0.11	0.11	0.16	0.11	0.17
BaO																	
CO_2																	
Ba	450	500	600	600	700	450	400	300	500	700	600	400	600	400	500	450	500
Sr	800	1200	1200	800	900	1200	1100	840	700	700	700	810	860	700	600	740	960
Rb	20	20	20	<10	20	20	20	20	20	10	10	10	10	<10	20	20	20
Zr	70	50	60	60	60	20	50	60	50	60	50	60	50	50	50	50	50
Y																	
La																	
Ce																	
Nd																	
Sm																	
Nb	22	26	17	24	19	22	23	20	22	17	37	24	27	22	18	20	22
Li	8.6	6.1	9.4	9.0	8.1	7.0	6.1	9.4	7.7	9.8	8.0	7.7	8.3	8.5	9.0	9.0	9.0
B																	
U																	
Th																	
Mo	1.5	1.0	1.0	0.8	0.8	0.5	0.8	1.5	1.5	1.0	1.0	1.5	1.0	1.5	1.0	2.0	0.8
Pb	7	6	5	6	5	3	5	5	6	25	6	8	6	7	6	7	4
Zn	140	140	130	140	140	140	140	150	150	150	140	140	150	140	130	140	140
Cu	30	35	33	32	32	38	34	35	35	38	34	34	40	33	34	41	34
Ni	74	78	100	86	93	150	100	70	82	74	82	64	74	90	82	110	86
Cr	200	250	200	180	150	200	200	180	150	180	200	200	200	200	200	200	200
3	150	200	200	180	200	100	150	150	150	150	170	200	200	200	150	200	180
Co	50	34	42	42	46	42	46	38	50	46	50	42	46	38	42	46	44
Sn	3	3	3	3	3	2	2	3	3	5	3	3	3	4	3	3	3
Ta	1.0	0.8	0.9	0.9	1.0	1.0	1.0	1.1	0.9	0.8	0.9	0.9	1.2	1.1	0.9	0.8	1.0
Ga	15	17	20	20	18	18	17	17	15	15	15	17	15	17	15	20	17
$^{87}Sr/^{86}Sr$																	
$^{143}Nd/^{144}Nd$																	
AGE																	
Reference	30	30	30	30	30	30	30	30	30	30	30	30	30	30	30	30	30

APPENDIX 7 (continued)

	Locality 165 NE Tuva, N Field		Locality 165 NE Tuva, S Field											Locality 158-169 Sayan-Baikal Highlands		
SiO_2	49.90	50.12	43.82	45.74	46.40	47.10	47.20	48.07	49.00	49.12	49.15	49.75	48.63	47.00	47.86	48.81
Al_2O_3	14.16	16.64	14.21	16.82	13.00	16.36	14.97	13.90	15.51	15.99	13.15	15.19	17.42	16.24	15.90	16.55
Fe_2O_3	4.27	4.23	3.91	2.81	5.00	3.99	4.62	4.22	3.05	2.57	6.10	4.43	4.28	5.35	5.33	5.35
FeO	6.10	5.24	6.97	7.40	7.54	6.61	5.75	6.82	6.75	7.18	5.75	6.10	6.53	7.07	7.50	6.51
MgO	6.81	6.45	10.55	8.99	7.17	6.97	7.15	7.12	6.85	8.50	7.13	6.30	6.50	6.97	8.19	5.94
CaO	8.04	7.18	8.14	7.97	8.86	8.19	8.09	8.43	8.02	6.85	7.74	6.86	6.50	8.49	8.85	8.01
Na_2O	3.76	3.40	3.58	3.70	3.22	2.17	3.18	3.18	2.74	2.74	3.20	3.62	2.90	3.74	2.21	3.23
K_2O	2.40	2.20	1.13	2.20	2.40	1.63	2.40	2.27	2.15	1.60	2.14	2.26	2.28	2.04	1.08	2.17
H_2O+	0.89	1.06	2.56	0.80	2.00	2.22	2.93	1.89	1.64	0.90	1.92	1.92	0.28			
H_2O-	0.52	0.42	0.62	0.46	0.40	0.64	0.68	0.72	0.46	0.48	0.73	0.63	0.59			
TiO_2	2.33	2.22	2.30	2.25	2.38	2.55	2.12	2.07	2.07	2.14	2.13	2.02	2.21	2.24	2.26	2.34
P_2O_5	0.74	0.74	0.93	0.77	0.90	0.82	0.70	0.74	0.70	0.72	0.70	0.74		0.66	0.57	0.81
MnO	0.18	0.10	0.07	0.12	0.19	0.11	0.18	0.19	0.11	0.11	0.17	0.17	0.08	0.20	0.25	0.28
BaO																
CO_2																
Ba	400	700	700	400	400	400	400	600	400	500	400	300				
Sr	700	700	1000	900	860	940	800	900	800	800	700	900				
Rb	20	10	10	20	20	20	10	20	10	20	10	10				
Zr	70	50	50	40	50	60	60	40	40	40	50	50				
Y																
La																
Ce																
Nd																
Sm																
Nb	23	20	39	26	37	32	25	26	23	25	25	29				
Li	9.0	7.3	8.1	7.8	7.3	8.6	7.8	7.8	7.8	9.0	8.1	9.0				
B																
U																
Th																
Mo	1.5	1.5	1.5	2.0	0.8	1.5	1.0	1.5	0.8	0.8	1.0	2.0				
Pb	7	3	7	10	7	7	7	7	8	7	6	10				
Zn	140	140	160	140	140	150	130	130	160	150	150	150				
Cu	32	36	40	36	38	25	30	28	34	36	42	32				
Ni	74	96	86	98	82	60	64	74	78	78	100	82				
Cr	200	200	200	200	200	150	200	200	100	200	200	100				
V	170	200	200	200	200	200	200	150	150	200	180	120				
Co	42	50	44	48	46	50	50	50	52	46	46	42				
Sn	3	3	3	3	4	3	3	3	3	3	3	4				
Ta	0.9	0.9	1.4	1.2	1.6	1.2	1.0	0.8	1.0	0.9	1.3	0.9				
Ga	15	18	17	15	17	18	17	17	15	17	18	18				
$^{87}Sr/^{86}Sr$																
$^{143}Nd/^{144}Nd$																
AGE																
Reference	30	30	30	30	30	30	30	30	30	30	30	30	19	31	31	31

APPENDIX 7 (continued)

	Locality 158-169 Sayan-Baikal		Locality 170 Pyinmana
SiO2	47.13	46.23	47.96
Al2O3	16.23	15.01	1.96
Fe2O3	5.34	5.39	2.47
FeO	7.02	8.91	7.01
MgO	4.56	8.67	8.91
CaO	8.43	8.81	9.39
Na2O	3.06	2.68	3.24
K2O	1.76	1.26	1.39
H2O+		1.26	
H2O-		1.36	
TiO2	2.61	2.32	1.23
P2O5	0.60		0.14
MnO	0.24	0.15	0.16
BaO			
CO2			
Ba			
Sr			
Rb			
Zr			
Y			
La			
Ce			
Nd			
Sm			
Nb			
Li			
B			
U			
Th			
Mo			
Pb			
Zn			
Cu			
Ni			
Cr			
V			
Co			
Sn			
Ta			
Ga			
87Sr/86Sr			
143Nd/144Nd			
AGE			
Reference	31	19	11

Notes

References to geochemical data:

1. Tomita (1969)
2. Lee (1981)
3. Tsung-pu (1960)
4. Ogura (1969)
5. Cong Bolin and others (1979)
6. Zhou and Armstrong (1982)
7. Wang (1974)
8. Zhu and others (1983)
9. Deng Wanming (1978a)
10. Chhibber (1934)
11. Stephenson and Marshall (1984)
12. Bender (1983)
13. Gobbett and Hutchison (1973)
14. Grubb (1965)
15. Barr and Macdonald (1978)
16. Barr and Macdonald (1981)
17. Barr and Macdonald (1979)
18. Lacombe (1969)
19. Kepezhinskas and Luchitskiy (1972)
20. Kiselev and others (1978)
21. Kepezhinskas and others (1975)
22. Filippov and others (1976)
23. Gorodinskiy and others (1967)
24. Vlodavetz and Piip (1959)
25. Bersenev and others (1966)
26. Stupak (1978a)
27. Stupak (1978b)
28. Bagin and others (1971)
29. Gerasimovsky and others (1980)
30. Gladkikh and others (1980)
31. Belov (1958)

Abbreviations:

ET = Early Tertiar
PAL = Paleocene
NE = Neogene
H = Nolocene
LT = Late Tertiary
MI = Miocene
PLE = Pleistocene
PRE-H = Prehistoric
LC = Late Cenozoic
PLI = Pliocene
Q = Quaternary
Hist = Historic

APPENDIX 8. CATALOG NUMBERS OF VOLCANIC AREAS

Ali-nazar Qurghan: 50
Altai Range: 124
Aluchin Group: 127
Anadyr River (Upper): 126
Anadyr Uplands: 135
Anyuy Group: 128
Arlykgol'sk: 169
Arts-Bogdo Range: 124

Bamaoqiongzong: 50
Ban Chay: 100
Ban Chiang Khian: 73
Ban Huai Sai: 85
Barun-Khobol Basin: 169
Bar Yakkar: 91
Bayan-leg Somon: 124
Beijing: 35
Bokeo: 86
Bol Anyuy: 135
Bolovens Plateau: 92
Bol'shoy Yenisey: 168
Bo Phloi: 77
Buriram: 81

Caoshi: 11
Cardamomes: 91
Cham: 90
Chamar-Daban: 159
Chanthaburi: 84
Chaun Bay: 135
Chekiang (E.): 44
Chiang Khong: 85
Ch'i-hsia: 41
Ch'ih-hsing-shan Group: 10
Chika: 169
Chi-shan: 43
Chindwin, Lower: 63
Ching-pe: 13
Chingpohu: 13
Ch'uga-ryong: 3
Chulutu: 122
Chup: 89

Da-Gang: 43
Dai-hang: 28
Dalinor: 29
Dariganga: 125
Darlac Plateau: 101
Datong: 36
Datun: 8
Denchai: 76
Dgida Basin: 158
Djiring Plateau: 101
Dolon Nor: 28
Dong Nai, Bas: 107
Dong Nai, Haut: 103
Dunhua: 12
Dzhida Basin: 158
Dzuunbayan: 109

Elovsky Spur: 160, 161
Erh-ko-shan: 19
Erkeshan: 19

Fanjiatun: 8
Fei-Hsien: 39
Feixian: 39
Fuxin: 40

Gacong: 23

APPENDIX 9. CATALOG NUMBERS OF NAMED VOLCANOES

Aghie (Peishan): 53*
Akhalik: 161
Akhat: 125
Aku: 155
Alpha: 127
Aniusk: 128*
Ap-San: 3

Baegdu (P'aektu-san): 5*
Baekdoosan (P'aektu-san): 5*
Baerqian: 27
Baevsky: 150
Baishan: 53*
Baitoushan (P'aektu-san): 5*
Balagan Tas: 132*
Banbanlongwan (Pan-pan-lung-wan): 6
Barun-nerte-ula: 125
Bayan-tsagan: 125
Bayan-tsagan-dzurikhe: 114
Beigelaqiushan (Pei-ko-la-chiu): 20
Beta: 127
Bichbalick (Peishan): 53*
Bijiashan (Pi-chian): 20
Bilibina: 129
Bogo-togo-ula: 115
Bo Kham: 86
Bolonsky: 142
Boschan (Peishan): 53*
Bu-Jerman: 102
Bujishan: 36
Bu Kong Rong: 103
Bus Obo: 114

Cam Tiem: 106
Changbaishan (Chang-pai): 5
Changgun-bong: 5
Chang-pai-shan (Changbaishan): 5
Chengzilou: 49
Chepe: 154
Chia-ts'ung (Jiacong): 23
Chien-shai-ting-tzu: 6
Chien-shan (Jianshan): 8
Chien-shan (Jianshan): 13
Chi-Haedrong: 99
Chi-Holam: 99
Ching-i-shan (Jingyishan): 7
Ching-lung-ting-tzu (Jinglongdingzi ?): 6
Chi-ten: 99
Ch'on-ji: 5
Cu-lao Bo Bai: 96
Cu-lao Re Island: 97

Dabingshan (Ta-ping): 15
Dadongbinglingshan (Ta-tun-ping-ling): 8
Dagushan (Ta-ku-): 9
Dagushanzi (Ta-ku-shan-tsu): 6
Daheishan: 22
Dakongshan: 49
Dalong: 49
Dalongwanshan (Ta-lung-wan): 6
Datuanshan: 49
Datugeerjishan (Ta-tu-ko-erh-chi): 10
Datungdongshan (Ta-t'un-tung-): 8
Dayenzhifeng (Ta-yen-chih-feng): 5
Dayingshan (Tay-in): 49*
Dayishan (Ta-i-): 6
Debussey: 50
Derbi-Tayga: 168
Dombrovskiy: 157
Donghalabanshan (Tung-ha-la-pa-): 10
Donghengdaoshan (Tung-heng-tao-): 6
Dongjianshan (Tung-chien-): 9
Dongjiaodebushan (Tung-chiao-te-pu-): 20
Donglongwan (Tung-lung-wan): 6
Donglongmenshan (Tung-lung-men-): 20
Duanzishan: 22
Dulan-Khara: 114
Durbulzhi-Ula: 120
Dushu-sharo-ula: 123
Dusy: 125
Dvoinoi: 153
D'yakhtardakh: 134
Dzan Tologoi: 122
Dzhiran: 124
Dzodal-khan: 125
Dzun-nerete: 125

Elgytkhyn: 135†
Erh-lung-wan: 6
Erlongwan (Erh-lung-wan): 6
Eshougedashan: 36

Fangshan: 43

Gacong (Ka-tsung): 23
Gamma: 127*
Gaoshanling: 47
Gremuchaya: 126

Halunershan (Baerqian): 27
Han Bogdo Ula: 109
Hawshuenshan (Maanshan): 49
Heikongshan: 49
Heishan: 36
Henwoshan: 36
Hill 5891: 54
Hochan (Peishan): 53*
Ho-shuen-shan (Maanshan): 49
Hotcheou (Turfan): 51*
Hsiao-i-shan (Xiaoyishan): 6
Hsiao-i-shan: 22
Hsiao-ko-shan (Xiaokeshan): 19
Hsiao-ku-shan (Xiaogushan): 9
Hsiao-ku-shan: 6
Hsiao-lung-wan (Xiaolongwan): 6
Hsiao-ta-ko-erh-chi-shan (Xiaotugeerjishan): 10
Hsiao-weng-chuan (Xiaolognwan ?): 6
Hsiao-yen-chi-feng (Xiaoyanzhifeng): 5
Hsi-chiao-te-pu-shan (Xijiaodebushan): 20
Hsi-chien-shan (Xijianshan): 9
Hsi-ha-la-pa-shan (Xihalabashan): 10
Hsi-heng-tao-shan (Xihengdaoshan): 6
Hsi-lung-men-shan (Xilongmenshan): 20
Hsing-an-nan-shan (Xingannanshan ?): 6
Hsi-pai-tzu-lung-wan (Xipaizilongwan): 6
Hsi-shan (Xishan): **19**
Hsi-shan: 22
Huang-i-shan (Huangyishan): 7
Huangyishan (Huang-yi): 7
Huangzo: 47
Hululongwan (Hu-lu-lung-wan): 6
Hu-lu-lung-wan: 6
Huoshaoshan (Hu-shao-): 20*
Hwang-bong: 5

Ih-jia-shan (Yijiashan?): 21
Ika-togo-ula: 115
Imansky: 145
Inde Hill: 66
Indighirsky: 132*

Jiacong (Chia-ts'ung): 23
Jianshan (Chien-): 8
Jiaoshan: 49
Jiaoyishan: 49
Jingyishan (Ching-I-): 7
Jinlongdingzi (Chin-lung-ting-tzu): 6
Jinshan: 36
Jiujinshansi: 36
Jiyunling: 46

Kardas: 54*
Ka-tsung (Gacong): 23
Keluonanshan (K'o-lo-nan-): 22
Khala (Peishan): 53*
Khao Phloi Waen: 84
Khara-Boldock: 160
Khobok: 161
Kholbo-teg-ula: 123
Khorog: 122
Khulugayshi: 162
Khutza-ula: 109
K'o-lo-nan-shan (Keluonanshan): 22
Kovrizhka: 160
Kropotkin: 163
Kuan-tzo-ling: 6
Kuntinsk: 161
Kwandu-bong: 5
Kyrgannivaam: 126

La Bang: 99
Laifengshan: 49
Langwoshan: 36
Laoguipo (Lao-kuei-po): 49
Laoheidingzi (Changpaishan): 5
Laoheishan (Lao-hei-): 20*
Lao-hei-ting-tzu (Changpaishan): 5
Lao-kuei-po (Laoguibo): 49
Leihufeng: 47
Leshe Craters: 63
Lien Khang: 104
Lingshi: 46
Lom Kot: 86
Longtang: 47

Yaowushan: 49
Ya-pa-kang (Yabagang): 6
Ya Ra: 86
Yen-pei-shan (Yanbeishan): 20
Yijiashan (Ih-chia): 21
Yiongzing: 47
Ywatha: 63

Key: *Volcanoes reported as historically active; † meteorite crater; †? reported as active but probably not.

APPENDIX 10. UNNAMED VOLCANOES

Catalog Number	Number of Volcanoes	Catalog Number	Number of Volcanoes	Catalog Number	Number of Volcanoes
3	1	46	9	118	5
5	100+	47	7	119	5
6	11	49	45?	122	4
6	20	50	3+ *	125	222
7	2	54	1+ *	126	25
16	2?	56	5	128	1
17	2	74	1+	130	1
20	2	76	1+	131	1
22	1	86	1+	134	?
23	1	93	4	139	5
25	6	94	3	153	5
29	30	100	2+	157	?
31	9	109	18	158	?
32	2	111	3	160	12
33	3	112	3	164	2
34	12	113	2	165	11
36	10	116	10	166	?
37	1	117	2	170	1

*Volcanoes reported as historically active.

REFERENCES CITED

Allen, C. C., 1980, Volcano-ice interactions on the earth and Mars: Advances in Planetary Geology, NASA TM-81979, p. 163–264.

Alyunin, A. V., and Dabhiza, A. I., 1980, Geophysical characteristics of meteoric crater [abs.]: Houston, The Lunar and Planetary Institute, Lunar and Planetary Science, XI, p. 21–23.

Anderson, J. G., 1923, Essays on the Cenozoic of Northern China: Memoirs of the Geological Survey of China, series A, no. 3.

Andrews-Speed, C. P., 1981, The case against a Phanerozoic Kolyma plate in the northeastern U.S.S.R.: Geology, v. 9, p. 174–177.

Argunov, M. S., and Gavrikov, S. I., 1960, Balagan-Tas, an early Quaternary volcano: Izvestiya Akademiia Nauk SSSR Series Geologiya, v. 8, p. 72–75.

Bagin, V. I., Brodskaya, S. Yu., Petrova, G. N., Pecherskiy, D. M., and Saltykovskiy, A. Ya., 1971, Certain distinctions of basaltic volcanism in Vitim Plateau: International Geology Review, v. 14, p. 225–229.

Barbour, G. B., 1929, The geology of the Kalgan area: Memoirs of the Geological Survey of China, series A, no. 6.

Barr, S. M., and Macdonald, A. S., 1978, Geochemistry and petrogenesis of late Cenozoic alkaline basalts of Thailand: Geological Society of Malaysia Bulletin, v. 10, p. 25–52.

——, 1979, Paleomagnetism, age, and geochemistry of the Denchai Basalt, northern Thailand: Earth and Planetary Science Letters, v. 46, p. 113–124.

——, 1981, Geochemistry and geochronology of late Cenozoic basalts of Southeast Asia: Geological Society of America Bulletin, v. 92, part I, p. 508–512, part II, p. 1069–1142.

Barr, S. M., Macdonald, A. S., Haile, N. S., and Reynolds, P. H., 1976, Paleomagnetism and age of the Lampang Basalt (northern Thailand), and age of the underlying pebble tools: Journal of the Geological Society of Thailand, v. 2, p. 1–10.

Basaltic Volcanism Study Project, 1981, Basaltic volcanism on the terrestrial planets: New York, Pergamon Press, 1286 p.

Beckinsale, R. D., Suensilpong, S., Nakapadungrat, S., and Walsh, J. N., 1979, Geochronology and geochemistry of granite magmatism in Thailand in relation to a plate tectonic model: Journal of the Geological Society of London, v. 136, p. 529–537.

Belov, I. V., 1958, Facies and chemical composition of the trachybasalts of the Sayan-Baikal highlands: Izvestiya Akademiia Nauk SSSR Series Geologiya, v. 7, p. 80–91.

Ben-Avraham, Z., and Uyeda, S., 1973, The evolution of the China basin and the

Mesozoic paleogeography of Burma: Earth and Planetary Science Letters, v. 18, p. 365–376.

Bender, F., 1983, Geology of Burma: Berlin-Stuttgart, Gebruder Borntraeger, 293 p.

Berkey, C. P., and Morris, F. K., 1927, Geology of Mongolia; Natural history of central Asia: New York, American Museum of Natural History, 475 p.

Bersenev, I. I., Buryanova, I. Z., Kasyan, E. D., and Licht, F. R., 1966, Tufflavas of southern Sikhote-Alin, *in* Cook, E. F., ed., Tufflavas and ignimbrites: New York, Elsevier, p. 112–118.

Bonney, T. G., 1899, Volcanoes; Their structure and significance: New York, G. P. Putnam's Sons, 332 p.

Brown, G. F., Buravas, S., Charaljavanaphet, J., Jalichandra, N., Johnston, W. D., Sresthaputra, V., and Taylor, G. G., 1951, Geological reconnaissance of the mineral deposits of Thailand: U.S. Geological Survey Bulletin 948, 183 p.

Burrett, C. F., 1974, Plate tectonics and the fusion of Asia: Earth and Planetary Science Letters, v. 21, p. 181–189.

Burton, R. C., 1913, Contributions to the geology of the province of Yunnan in western China (II); Petrology of the volcanic rocks of the Teng-Yueh District: Records of the Geological Survey of India, v. 43, p. 206–227.

Carbonnel, J. P., Duplaix, S., and Selo, M., 1972, La méthode des traces de fission de l'uranium appliquée a la géochronologie. Datation du magmatisme recent de l'Asie du sud-est: Revue de Géographie Physique et de Géologie Dynamique, v. 14, p. 29–46.

Carbonnel, J. P., Selo, M., and Poupeau, G., 1973, Fission track age of the gem deposit of Pailin (Cambodia) and recent tectonics in the Indochinan province: Modern Geology, v. 4, p. 61–64.

Chandra, U., 1979, Large-scale Cenozoic tectonics of central and south-central Asia; Products of continental collision: Physics of Earth and Planetary Interiors, v. 20, p. 33–41.

Chasovitin, M. D., 1963, New information on the intracontinental Quaternary volcanoes of northeastern Asia: Doklady Akademiia Nauk SSSR, v. 152, p. 39–41.

Chhibber, M. D., 1934, Geology of Burma: London, Macmillan, 538 p.

China Pictorial, 1980, part 10, p. 8–12.

Chinese Earthquakes, 1981: Lishan Bangzhu, p. 336–342 (in Chinese).

Chingchang, B., 1979, Sinokorea-Cathaysia in its Mesozoic-Cenozoic convergence zone setting: Memoirs of the Geological Society of China, v. 3, p. 73–79.

Coggin Brown, J., 1913, Contributions to the geology of the province of Yunnan in western China (I); The Bhamo–Teng-Yueh area: Records of the Geological Survey of India, v. 43, p. 174–205.

Cong Bolin, Zhang Wenhua, and Ye Danian, 1979, The study on the Cenozoic basalts in north China fault block: Acta Geologica Sinica, v. 53, p. 114–124 (in Chinese).

Crawford, A. J., Beccaluva, L., and Serri, G., 1981, Tectonomagnetic evolution of the West Philippine–Mariana region and the origin of boninites: Earth and Planetary Science Letters, v. 54, p. 346–356.

Curray, J. R., Moore, D. G., Lawver, L. A., Emmel, F. J., Raitt, R. W., Henry, M., and Kieckhefer, R., 1979, Tectonics of the Andaman Sea and Burma, *in* Watkins, J. S., Montadert, L., and Dickerson, P., eds., Geological and geophysical investigations of continental margins: Memoirs of the American Association of Petroleum Geologists 29, p. 189–198.

Das, S., and Filson, J. R., 1975, On the tectonics of Asia: Earth and Planetary Science Letters, v. 28, p. 241–253.

Deng Wanming, 1978a, A preliminary study on the petrology and geochemistry of the Quaternary rocks of northern Tibet Autonomous Region: Acta Geologica Sinica, v. 52, p. 148–162 (in Chinese).

——, 1978b, Volcanoes on the Chang-tang: China Pictorial, Part 1, p. 10–11.

Devyatkin, Ye. V., and Smelov, S. B., 1979, Position of basalts in the Cenozoic sedimentary sequence of Mongolia: International Geology Reviews, v. 22, p. 307–317.

E Molan, Deng Jinfu, and Lu Fengxiang, 1981, Melting events of the upper mantle in Wangqing area, Jilin Province, China [abs.]: 1981 International Association of Volcanology and Chemistry of the Earth's Interior Symposium, Tokyo-Hakone, p. 209–210.

Eguchi, T., Uyeda, A., and Maki, T., 1979, Seismotectonics and tectonic history of the Andaman Sea: Tectonophysics, v. 57, p. 35–51.

Feng Maoseng, and Whitford-Stark, J. L., 1986, The 1719–21 eruptions of potassium-rich lavas at Wudalianchi, China: Journal of Volcanology and Geothermal Research, v. 30 (in press).

Feng Maoseng, Guo Keyi, and Wang Fuquan, 1979, Wudalianchi Volcanoes in China: Shanghai, Shanghai Scientific and Technical Publishers, 85 p. (in Chinese with English introduction).

Filippov, L. V., Lipovskiy, Yu. O., and Kapitonova, A. Ya., 1976, Central Mongolian potassic basaltoids and abyssal magma formation: Geochemistry International, v. 13, p. 106–118.

Florensov, N. A., and Logatchev, N. A., 1975, Rifting problems; Symposium on rift zones of the Earth; Field Excursion: Irkutsk, U.S.S.R. Akademiya Science, p. 124–131.

Florensov, N. A., and Solonenko, V. P., 1963, The Gobi-Altai Earthquake: Moskva, Izdatel'stvo Akademii Nauk SSSR, 424 p.

Fujita, K., and Newberry, T., 1983, Accretionary terranes and tectonic evolution of northeast Siberia, *in* Hashimoto, M., and Uyeda, S., eds., Accretionary tectonics in the circum-Pacific regions: Tokyo, Terra Scientific Publishing Company, p. 43–57.

Genshaft, Yu. S., Zaytsev, N. S., and Saltykovskiy, A. Ya., 1978, Construction of the deep-seated zones and basaltic volcanism of Mongolia: Geotectonics, v. 12, p. 392–399.

Gerasimovsky, V. I., Roshchina, I. A., and Shevaleyevskiy, I. D., 1980, Chemical compositions of basalts in the Baykal rift zone: Geochemistry International, v. 17, p. 1–10.

Gladkikh, V. S., Konova, N. I., Solov'yev, V. A., and Altukhov, Ye. N., 1980, Petrochemical and geochemical characteristics of undifferentiated alkali olivine basalts of northeast Tuva: Geochemistry International, v. 17, p. 126–133.

Gobbett, D. J., and Hutchison, C. S., 1973, Geology of the Malay Peninsula: New York, Wiley Interscience, 438 p.

Gorodinskiy, M. Ye., Dovgal, Yu. M., and Sterligova, V. Ye., 1967, The Aluchin group of Late Quaternary volcanoes in western Chukotka: International Geology Reviews, v. 10, p. 1045–1054.

Grenard, F., 1929, Géographie Universelle, Tome VIII, Haute Asie: Paris, Librarie Armand Colin, p. 235–394.

Grubb, P.L.C., 1965, Undersaturated potassic lavas and hyperbyssal intrusives in north Johore: Geological Magazine, v. 102, p. 338–346.

Gurov, E. P., Valter, A. A., Gurova, E. P., and Kotloskaya, F. I., 1979, Elgygytgyn impact crater, Chucotka; Shock metamorphism of volcanic rocks [abs.]: Houston, Lunar and Planetary Institute, Lunar and Planetary Science X, p. 479–481.

Herman, B. M., Anderson, R. N., and Truchan, M., 1979, Extensional tectonics in the Okinawa Trough, *in* Watkins, J. S., Montadert, L., and Dickerson, P., eds., Geological and geophysical investigations of continental margins: Memoirs of the American Association of Petroleum Geologists 29, p. 199–208.

Hsieh, C. Y., 1963, Major geotectonic features of southeast China: Scientica Sinica, v. 12, p. 1871–1887.

Humboldt, A. Von, 1830, Sur les Volcans de l'Asie centrale: Annales de Chemie et de Physique Series 2, v. 45, p. 337–348.

Huzita, K., 1984, Fault block movements in Shandong Province, China, *in* Uemura, T., and Mizutani, S., eds., Geological structures: New York, John Wiley & Sons, p. 44–56.

IAVCEI, 1951–75, Catalog of active volcanoes of the world, v. 1–22, Rome, International Association of Volcanology and Chemistry of the Earth's Interior.

——, 1973, Data Sheets of the post-Miocene volcanoes of the world with index maps: Rome, International Association of Volcanology and Chemistry of the Earth's Interior.

Jacobson, H. S., Pierson, C. T., Danusawad, T., Japakasetr, T., Inthuputi, B.,

Siriratunamongkol, C., Prapassornkuls, S., and Pholphan, N., 1969, Mineral investigations in northeast Thailand: U.S. Geological Survey Professional Paper 618, 96 p.

Jin Chengwei, 1981, The Mesozoic-Cenozoic volcanic rocks in Xizang and their bearing on geology, *in* Geological and Ecological Studies of Qinghai-Xizang Plateau: New York, Gordon and Breach, p. 433–442.

Jurdy, D. M., 1979, Relative plate motions and the formation of marginal basins: Journal of Geophysical Research, v. 84, p. 6796–6802.

Karig, D. E., 1971, Origin and development of marginal basins in the western Pacific: Journal of Geophysical Research, v. 76, p. 2542–2561.

Kepezhinskas, V. V., and Luchitskiy, I. V., 1972, Petrochemistry of Cenozoic volcanic rocks of central Asia: Doklady Akademiya Nauk SSSR, v. 205, p. 59–62.

Kepezhinskas, V. V., Devyatkin, E. V., and Dashdava, Z., 1975, Cenozoic basaltoids of the Taryatu Depression (Mongolia): Geologiya i Geofizika, v. 16, p. 3–14.

Kidd, W.S.F., 1973, Supplement to Catalog of Active Volcanoes of the World: Bulletin Volcanologique, p. 48–49.

Kim, S-Y., 1979, Analyses of lineaments extracted from LANDSAT images of the Korean Peninsula: Nagoya University, Journal of Earth Sciences, v. 26/27, p. 49–74.

Kiselev, A. I., Golovko, H. A., and Medvedev, M. E., 1975, Chemistry and petrogenesis of the basalts from the Baikal rift zone, *in* Rifting problems; Symposium on rift zones of the Earth: Irkutsk, U.S.S.R. Akademiya Science, p. 44–46.

—— , 1978, Petrochemistry of Cenozoic basalts and associated rocks in the Baikal rift zone: Tectonophysics, v. 45, p. 49–59.

Klimetz, M. P., 1983, Speculations on the Mesozoic plate tectonic evolution of eastern China: Tectonics, v. 2, p. 139–166.

Kobayashi, T., 1956, A contribution to the geo-tectonics of North Korea and South Manchuria: Journal of the Faculty of Science, University of Tokyo, Section II, v. 10, p. 133–311.

Krasnyy, L. I., 1967, Geoblocks: Geotectonics, v. 1, p. 318–326.

Kuno, H., 1953, Plateau basalt lavas of eastern Manchuria: Proceedings of the 7th Pacific Science Congress, 1949, v. 2, p. 375–382.

Lacombe, P., 1969, Le massif basaltique Quaternaire à ziron-gemmes de Ratanakiri (Cambodge nord-oriental), premiere partie: Bulletin du Researches Géologiques et Minières, Second Series, Section 4, no. 3, p. 31–91.

Lallemand, S., and Jolivet, L., 1986, Japan Sea: A pull-apart basin?: Earth and Planetary Science Letters, v. 76, p. 375–389.

La Moreaux, P. E., Charaljavanaphet, J., Jalichan, N., Chiengmai, P.P.N., Bunnag, D., Thavisri, A., and Rakprathum, C., 1958, Reconnaissance of the geology and groundwater of the Khorat Plateau, Thailand: U.S. Geological Survey Water-Supply Paper 1429, 63 p.

Le Dain, A. Y., Tapponnier, P., and Molnar, P., 1984, Active faulting and tectonics of Burma and surrounding regions: Journal of Geophysical Research, v. 89, p. 453–472.

Lee, D. S., 1981, Volcanic activity in the southwest part of the Choogaryong rift valley, Korea [abs.]: 1981, International Association of Volcanology and Chemistry of the Earth's Interior symposium, arc volcanism, Tokyo-Hakone, p. 200–202.

Lee, J. S., 1939, The geology of China: London, T. Murby and Co., 528 p.

Li, W.Y.L., 1970 Geologic map of the Republic of China: Chinese Petroleum Corporation, scale 1:4,000,000.

Lu Fengxiang, Deng Jinfu, and E Molan, 1981, On the origin of the alkaline basaltic magma of the Huangyishan volcanic cone, Kuandian County, Liaoning Province, China [abs.]: 1981, International Association of Volcanology and Chemistry of the Earth's Interior arc symposium, Tokyo-Hakone, p. 209–210.

Matsuda, T., and Uyeda, S., 1971, On the Pacific-type orogeny and its model; Extension of the paired-belts concept and possible origins of marginal seas: Tectonophysics, v. 11, p. 5–27.

Middlemost, E.A.K., 1972, A simple classification of volcanic rocks: Bulletin Volcanologique, v. 36, p. 382–397.

Mitchell, A.H.G., 1981, Phanerozoic plate boundaries in mainland SE Asia, the Himalayas, and Tibet: Journal of the Geological Society of London, v. 138, p. 109–122.

Mitchell, A.H.G., and McKerrow, W. S., 1975, Analogous evolution of the Burma orogen and the Scottish Caledonides: Geological Society of America Bulletin, v. 86, p. 305–315.

Miyashiro, A., 1978, Nature of the alkalic volcanic rock series: Contributions to Mineralogy and Petrology, v. 66, p. 91–104.

Molnar, P., and Tapponnier, P., 1975, Cenozoic tectonics of Asia; Effects of a continental collision: Science, v. 189, p. 419–426.

Nalivkin, D. V., 1973, Geology of the U.S.S.R. (translated by Rast, N.): Toronto, University of Toronto Press, 855 p.

Natal'in, B. A., and Parfenov, L. M., 1983, Accretional and collisional eugosynclinal folded systems of the northwestern Pacific rim, *in* Hashimoto, M., and Uyeda, S., eds., Accretion tectonics in the circum-Pacific region: Tokyo, Terra Scientific Publishing Co., p. 59–68.

Ninkovich, D., and Hays, J. D., 1972, Mediterranean island arcs and origin of potash volcanoes: Earth and Planetary Science Letters, v. 16, p. 331–345.

Norin, E., 1979, The relationships between the Tibetan platform and the Tarim basin: Bulletin of the Geological Institute, University of Uppsala, New Series, v. 8, p. 17–34.

Ogura, T., 1969, Volcanoes in Manchuria, *in* Ogura, T., ed., Geology and mineralogy of the Far East: Tokyo, University of Tokyo Press, p. 373–413.

Parfenov, L. M., and Natal'in, B. A., 1979, Mesozoic-Cenozoic tectonic evolution of northeast Asia: Doklady Akademiya Nauk SSSR, v. 235, p. 89–91.

Parfenov, L. M., Natal'in, B. A., Voynova, I. P., and Popeko, L. I., 1981, Tectonic evolution of active continental margins along the northwestern margin of the Pacific Ocean: Geotectonics, v. 15, p. 54–67.

Plakhotnik, V. G., 1968, Pliocene domal volcanoes in the lower Amur region: International Geology Reviews, v. 10, p. 44–47.

Ridd, M. F., 1980, Possible Paleozoic drift of SE Asia and Triassic collision with China: Journal of the Geological Society of London, v. 137, p. 635–640.

Rodolfo, K. S., 1969, Bathymetry and marine geology of the Andaman Basin and tectonic implications for southeast Asia: Geological Society of America Bulletin, v. 80, p. 1203–1230.

Saurin, E., 1936, Sur la constitution géologique de la province du Phu-Yen (Sud Annam): Paris, Comptes Rendus Academie des Sciences, v. 202, p. 431–433.

—— , 1967, La néotectonique de l'Indochine: Revue de Géographie Physique et de Géologie Dynamique, v. 9, p. 143–151.

—— , 1968, Présence de zircons dans les basaltes de Xuan Loc (Sud Viêt-Nam): Comptes Rendus Sommaires de la Societe Géologique de France, v. 8, p. 281.

Sengor, A.M.C., 1984, The Cimmeride orogenic system and the tectonics of Eurasia: Geological Society of America Special Paper 195, 82 p.

Sengor, A.M.C., and Kidd, W.S.F., 1979, Post-collisional tectonics of the Turkish-Iranian Plateau and a comparison with Tibet: Tectonophysics, v. 55, p. 361–376.

Shen Yuanchao, and Wang Xinghua, 1983, Volcanic rocks in Jingbo Lake and its tectonic setting [abs.]: XVIII General Assembly, International Union of Geodesy and Geophysics, Hamburg 1983, Programme and Abstracts, v. 1, p. 573.

Short, N. M., Lowman, P. D., Freden, S. C., and Finch, W. A., 1976, Mission to Earth; LANDSAT Views the World: Washington, D.C., National Aeronautics and Space Administration, SP-360, 459 p.

Sillitoe, R. H., 1977, Metallogeny of an Andean-type continental margin in South Korea, *in* Talwani, M., and Pitman, W. C., III, eds., Island arcs, deep sea trenches, and back-arc basins: Washington, American Geophysical Union, p. 303–310.

Simkin, T., Siebert, L., McClelland, L., Bridge, D., Newhall, C., and Latter, J. H., 1981, Volcanoes of the World: Stroudsburgh, Pennsylvania, Hutchinson Ross Press, 233 p.

Speranskaya, I. M., 1966, Ignimbrites in the volcanic sequence of the northern coast of the Sea of Okhotsk and their origin, *in* Cook, E. F., ed., Tufflavas and ignimbrites: New York, Elsevier, p. 119–129.

Stephenson, D., and Marshall, T. R., 1984, The petrology and mineralogy of Mt. Popa Volcano and the nature of the late-Cenozoic Burma volcanic arc: Journal of the Geological Society of London, v. 141, p. 747–762.

Stupak, F. M., 1978a, Late Quaternary volcanoes of the Sakukan River Basin (Udokan Range, Northern Transbaikalia): Geologiya i Geofizika, v. 19, p. 87–94.

——, 1978b, Chemical evolution of Cenozoic lavas of the Baykal rift system (as exemplified by the Udokansk lava plateau): Geochemistry International, v. 15, p. 26–34.

Takahashi, M., 1983, Space-time distribution of Late Mesozoic to Early Cenozoic magmatism in east Asia and its tectonic implications, *in* Hashimoto, M., and Uyeda, S., eds., Accretion tectonics in the circum-Pacific regions: Tokyo, Terra Scientific Publishing Company, p. 69–88.

Tapponnier, P., and Molnar, P., 1977, Active faulting and tectonics in China: Journal of Geophysical Research, v. 82, p. 2905–2930.

Tapponnier, P., Peltzer, G., Le Dain, A. Y., Armijo, R., and Cobbold, P., 1982, Propogating extrusion tectonics in Asia; New insights from simple experiments with plasticine: Geology, v. 10, p. 611–616.

Tateiwa, I., 1960, Outline of the geology of Korea: International Geology Reviews, v. 2, p. 1053–1070.

Taylor, B., and Hayes, D. E., 1980, The tectonic evolution of the South China Basin, *in* Hayes, D. E., ed., The tectonics and evolution of southeast Asia seas and islands: Washington, D.C., American Geophysical Union Geophysical Monograph 23, p. 89–104.

Teilhard de Chardin, P., 1926, Etude geologique sur la region du Dalai-noor: Memoires de la Societe Geologique de France (Nouvelle Serie), no. 7, p. 1–53.

Terman, M. J., 1977, Cenozoic tectonics of east Asia, *in* Talwani, M., and Pitman, W. C., III, eds., Island arcs, deep sea trenches, and back-arc basins: Washington, D.C., American Geophysical Union, p. 468–470.

Tikhonov, V. I., 1977, Stratovolcanoes in Mongolia: Doklady Akademiya Nauk SSSR, v. 233, p. 53–55.

Tokyo Geological Society, 1929, Geological atlas of eastern Asia.

Tomita, T., 1935, On the chemical compositions of the Cenozoic alkaline suite of the circum-Japan Sea region: Journal of the Shanghai Science Institute, Section II, v. 1, p. 227–306.

——, 1969, Volcanic geology of the Cenozoic alkaline petrographic province of eastern Asia, *in* Ogura, T., ed., Geology and mineralogy of the Far East: Tokyo, University of Tokyo Press, p. 139–179.

Tsung-pu, C., 1960, Petrochemical study of the Cenozoic basaltic rocks in eastern China (Part 1 of 2): International Geology Reviews, v. 2, p. 196–217.

UNESCO, 1972, Geological map of Asia and the Far East, 2nd ed., scale 1:5,000,000.

Ustiev, E. K., 1959a, Ånuy volcano and Quaternary volcanism of northeast Asia: International Geology Reviews, v. 1, p. 85.

——, 1959b, Aniusky volcano and localization problems of Quaternary volcanism in northeastern Asia: Bulletin Volcanologique, v. 20, p. 155–171.

——, 1974, Problems of volcanism and plutonism; Volcano-plutonic formations: International Geology Reviews, v. 7, p. 1994–2016.

Uyeda, S., and Kanamori, H., 1979, Back-arc opening and the mode of subduction: Journal of Geophysical Research, v. 84, p. 1049–1061.

Vinogradov, A. P., ed., 1967, Atlas of the lithological-paleogeographical maps of the U.S.S.R., Volume IV, Paleogene, Neogene, and Quaternary, Grossheim, V. A., and Khain, V. E., eds.: Moskva, Ministry of Geology of the U.S.S.R.

Vlodavetz, V. I., and Piip, B. I., 1959, Catalog of active volcanoes of the World, Part VIII, Kamchatka and continental areas of Asia: Napoli, International Volcanological Association.

Wang, C., 1974, Geology of the Wudalianchi volcanoes, *in* Volcanic geology and multiple utilization of mineral resources in Wudalianchi, Dedu, Heilongjiang Province: Publication of the Committee of Sciences and Technology, Heilongjiang Province, China, p. 3–78 (in Chinese).

Washington, H. S., 1900, The composition of kulaite: Journal of Geology, v. 8, p. 610–620.

Watanabe, T., Langseth, M. G., and Anderson, R. N., 1977, Heat-flow in back-arc basins of the western Pacific, *in* Talwani, M., and Pitman, W. C., III, eds., Island arcs, deep sea trenches, and back-arc basins: Washington, D.C., American Geophysical Union, p. 137–161.

Wei Junchao, Zhou Zhenjie, and Liu Zigui, 1980, Geologic application of remote sensing in Tengchong, Yunnan, China. Proceedings of International Symposium of Environment, No. 14, p. 1891–1900.

Whitford-Stark, J. L., 1983, Cenozoic volcanic and petrochemical provinces of mainland Asia: Journal of Volcanology and Geothermal Research, v. 19, p. 193–222.

Wood, C. A., 1978, Morphometric evolution of composite volcanoes: Geophysical Research Letters, v. 5, p. 437–439.

——, 1980, Morphometric evolution of cinder cones: Journal of Volcanology and Geothermal Research, v. 7, p. 387–413.

Wu Liren, Zhai Mingzhe, and Zheng Xiangshen, 1983, Cenozoic volcanic rocks in east China [abs.]: XVIII General Assembly, IUGG, Hamburg, 1983, Programme and Abstracts, v. 1, p. 283.

Xie Guanghong, 1981, Temporal and spatial distribution of Cenozoic volcanism in eastern China [abs.]: 1981 International Association of Volcanology and Chemistry of the Earth's Interior symposium, arc volcanism, Tokyo-Hakone, p. 419–420.

Yang Qilong and Shen Minzi, 1981, Yanbajan geothermal field in Tibet and Tangchong geothermal area in Yannan Province, their relation to magma activity [abs.]: 1981 International Association of Volcanology and Chemistry of the Earth's Interior symposium, arc volcanism, Tokyo-Hakone, p. 427.

Ye, H., Shedlock, K. M., Hellinger, S. J., and Sclater, J. G., 1985, The North China basin; An example of a Cenozoic rifted intraplate basin: Tectonics, v. 4, p. 153–169.

Zhang, Z. M., Liou, J. G., and Coleman, R. G., 1984, An outline of the plate tectonics of China: Geological Society of America Bulletin, v. 95, p. 295–312.

Zheng Xiangshen, 1983, The Cenozoic potash volcanic rocks in north Heilonggiang Province, China [abs.]: XVIII General Assembly, IUGG, Hamburg, 1983, Programme and Abstracts, Volume 1, p. 284.

Zheng Xuezheng, 1981, On the Cenozoic basalt in east China [abs.]: 1981 International Association of Volcanology and Chemistry of the Earth's Interior symposium, arc volcanism, Tokyo-Hakone, p. 442–443.

Zhou, J., 1985, The timing of calc-alkaline magmatism in parts of the Alpine-Himalayan collision zone and its relevance to the interpretation of Caledonian magmatism: Journal of the Geological Society of London, v. 142, p. 309–317.

Zhou, X., and Armstrong, R. L., 1982, Cenozoic volcanic rocks of eastern China; Secular and geographic trends in chemistry and strontium isotopic composition: Earth and Planetary Science Letters, v. 58, p. 301–329.

Zhu, B-Q., Mao, C-X., Lugmair, G. W., and Macdougall, J. D., 1983, Isotopic and geochemical evidence for the origin of Plio-Pleistocene volcanic rocks near the Indo-Eurasian collisional margin at Tengchong, China: Earth and Planetary Science Letters, v. 65, p. 263–275.

Zhupakhin, E. N., Zhitetskii, A. A., and Skorikov, R. A., 1977, Properties of the geophysical fields and structures of the Anadyr segment of the Okhotsk-Chukchi volcanogenic belt (new data): Geologiya i Geofizika, v. 18, p. 114–122.

Zonenshayn, L. P., 1967, Tectonics of the folded regions of central Asia (the structural patterns of the geosynclinal regions): Geotectonics, v. 1, p. 356–365.

Zonenshayn, L. P., and Gorodnitskiy, A. M., 1977, Paleo-Mesozoic and Mesozoic reconstructions of the continents and oceans, Article 2, Late Paleozoic and Mesozoic constructions: Geotectonics, v. 11, p. 159–171.

Zonenshayn, L. P., and Savotsin, L. A., 1981, Geodynamics of the Baikal rift zone and plate tectonics of Asia: Tectonophysics, v. 76, p. 1–45.

Typeset by WESType Publishing Services, Inc., Boulder, Colorado
Printed in U.S.A. by Malloy Lithographing Inc., Ann Arbor, Michigan